EP Seventh Reader Workbook

I'm ____________________________________.

This is my workbook. My favorite books are

____

____

____

____

The vocabulary in this workbook is based on, and
used by permission of, Easy Peasy All-in-One Homeschool.
For EP's online curriculum visit www.allinonehomeschool.com

For more information visit www.puzzlefast.com

ISBN-13: 978-1505466812
ISBN-10: 1505466814

About This Workbook

This is an offline workbook of vocabulary puzzles and games for Easy Peasy All-in-One Homeschool's reading course for Level 7. We've modified and expanded upon the online activities available at the Easy Peasy All-in-One Homeschool website (www.allinonehomeschool.com) so that your child can work offline if desired. Whether you use the online or offline versions, or a combination of both, your child will enjoy these supplements to the Easy Peasy reading course.

How to Use This Workbook

This workbook is designed to be used as a complement to Easy Peasy's reading curriculum, either the online or offline version. It provides ample activities to help your child master the vocabulary words in Level 7. For any given lesson, use the Activity List to pick out an activity and have your child work on it. Here's our suggestion:

Use the worksheets with lesson numbers for the specified lessons when:

- New vocabulary words are first introduced.
- The online course or EP reader instructs to review the vocabulary words or to play an online vocabulary game.

Use the additional worksheets any time during the course when:

- Your child needs more practice on a specific vocabulary set.
- Your child wants extra activities just for fun.

If your child initially has difficulty remembering all the words, don't worry. The first activity for each vocabulary set is to review all the words and their meanings. The matching activities provided are another a great way to reinforce the meanings of the words in preparation for the more challenging exercises in the workbook.

The solutions to selected activities are included at the end of the workbook.

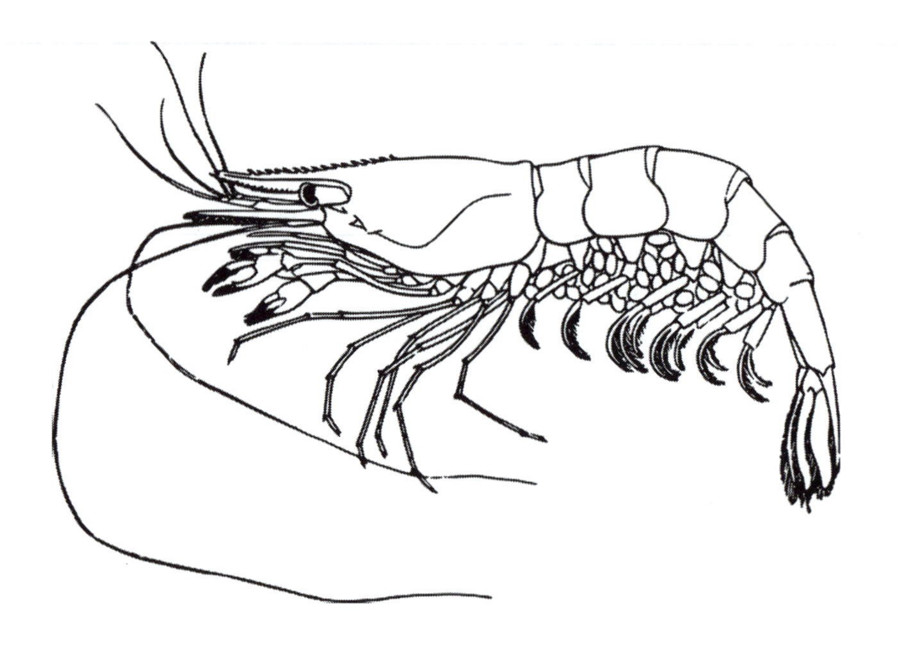

Activity List

Penrod and Sam I Vocabulary

Lesson 30

Review and read aloud the words and their meanings.

impudent = offensively bold; impertinent

broodings = depressing thoughts

resonant = resounding or echoing

plaintive = sounding sad and mournful

guttural = having a harsh, grating sound

barbaric = primitive; savagely cruel; exceedingly brutal

sonorous = capable of producing a deep or ringing sound

These broodings helped a little; but it was a severe morning, and on his way home at noon he did not recover heart enough to practice the bullfrog's croak, the craft that Sam Williams had lately mastered to inspiring perfection. This sonorous accomplishment Penrod had determined to make his own. At once guttural and resonant, impudent yet plaintive, with a barbaric twang like the plucked string of a Congo war-fiddle, the sound had fascinated him. It is made in the throat by processes utterly impossible to describe in human words, and no alphabet as yet produced by civilized man affords the symbols to vocalize it to the ear of imagination. "Gunk" is the poor makeshift that must be employed to indicate it.

– Booth Tarkington, *Penrod and Sam*

Penrod and Sam I Matching

Lesson 34

Can you match the words with their definitions?

broodings barbaric resonant

impudent GUTTURAL sonorous plaintive

________________ = primitive; savagely cruel; exceedingly brutal

________________ = capable of producing a deep or ringing sound

________________ = sounding sad and mournful

________________ = depressing thoughts

________________ = resounding or echoing

________________ = offensively bold; impertinent

________________ = having a harsh, grating sound

Penrod and Sam is a novel by Booth Tarkington that was first published in 1916. The book is the sequel to his 1914 work, *Penrod*, and focuses more on the relationship between the main character of the previous book, Penrod Schofield, and his best friend, Sam Williams. More of Penrod's adventures appear in the final book of the series *Penrod Jashber* (1929). The three books were published together in one volume, *Penrod: His Complete Story*, in 1931.

- Wikipedia, *Penrod and Sam*

Penrod and Sam II Vocabulary

Lesson 44

Review and read aloud the words and their meanings.

stratagem	= a plan or scheme
devoid	= entirely lacking or free from
unobtrusively	= inconspicuous, unassertive
truculence	= obstreperous and defiant aggressiveness
mollify	= appease the anger or anxiety of (someone)
plaintively	= expressing sorrow or melancholy; mournful
ferocity	= the state or quality of being ferocious; fierceness
dexterity	= skill in performing tasks, especially with the hands
converged	= come together from different directions so as eventually to meet
precociously	= unusually advanced or mature in development, especially mental development
grievance	= a real or imagined wrong or other cause for complaint or protest, especially unfair treatment

- ✓ She's always singing some old song plaintively.
- ✓ Eric's speech was devoid of warmth and feeling.
- ✓ Ambush is an effective military stratagem.
- ✓ The tiger attacked the antelope with ferocity.
- ✓ Will this be enough to mollify the outraged citizens?
- ✓ There was a signpost where the two paths converged.
- ✓ You need manual dexterity to be good at video games.
- ✓ Does your organization have a formal grievance procedure?
- ✓ From an early age she displayed a precocious talent for music.
- ✓ He spoke with the king quietly and unobtrusively for a while.

The Call of the Wild I Vocabulary

Lesson 46

Review and read aloud the words and their meanings.

latent = not shown

interlace = to intertwine

progeny = one's children

imperious = controlling; urgent

pervade = to spread everywhere

lacerate = to cut, especially by tearing

conveyance = the legal transfer of property

conciliate = to come together and be friends

obscure = unclear; to hide from view by covering

egotism = character trait of someone who loves to talk about himself

- ✓ Some viruses can remain latent in the body for years.
- ✓ Her hand was lacerated by a broken glass bottle.
- ✓ Red and blue ribbons were interlaced with one another.
- ✓ Conveyance of land is taxed in most states.
- ✓ The family lived obscurely and humbly in the shadow of the forest.
- ✓ An atmosphere of excitement pervaded the town.
- ✓ His egotism and arrogance annoyed everyone around him.
- ✓ The king was notorious for being imperious and having an explosive temper.
- ✓ After a long argument, Sally conciliated her brother by offering him some candy.
- ✓ The farmer worked hard and saved money so his progeny might have an easier life.

The Call of the Wild I Matching

Lesson 47

Can you match the words with their definitions?

progeny   egotism   pervade

interlace   imperious   conciliate   obscure

CONVEYANCE   latent   lacerate

______________ = unclear; to hide from view by covering

______________ = the legal transfer of property

______________ = to spread everywhere

______________ = one's children

______________ = not shown

______________ = to intertwine

______________ = controlling; urgent

______________ = to cut, especially by tearing

______________ = to come together and be friends

______________ = character trait of someone who loves to talk about himself

The Call of the Wild II Vocabulary

Lesson 48

Review and read aloud the words and their meanings.

courier = a messenger

arduous = difficult

belligerent = aggressive

consternation = sudden confusion or amazement

cadence = rhythmic movement or series of sounds

indiscretion = character trait of revealing private information

fastidious = delicate or refined; hard to please in matters of taste

vicarious = serving in the place of someone or something else

malinger = to pretend injury or sickness to avoid responsibility

retrogression = a reversal in development from a higher to lower state

✓ Retrogression to a child-like state is possible with some illnesses.
✓ The chef is very fastidious about the food he makes.
✓ The informant did not expect the detective's indiscretion.
✓ He gets a vicarious thrill out of watching adventure movies.
✓ During the game, the cheerleaders danced in a rapid cadence.
✓ After a long and arduous day's work, he was completely exhausted.
✓ She tried to calm him down but he remained upset and belligerent.
✓ The CEO's shocking decision caused consternation in the company.
✓ The doctor accused Tommy of malingering, but he was really sick.
✓ A courier brought the soldiers' letters from the battleground to the city.

The Call of the Wild II Matching

Lesson 49

Can you match the words with their definitions?

cadence belligerent indiscretion
fastidious courier arduous malinger
retrogression CONSTERNATION vicarious

________________ = character trait of revealing private information

________________ = sudden confusion or amazement

________________ = a messenger

________________ = aggressive

________________ = difficult

________________ = rhythmic movement or series of sounds

________________ = serving in the place of someone or something else

________________ = delicate or refined; hard to please in matters of taste

________________ = a reversal in development from a higher to lower state

________________ = to pretend injury or sickness to avoid responsibility

The Call of the Wild III Vocabulary

Lesson 54

Review and read aloud the words and their meanings.

innocuous = harmless

salient = conspicuous

inexorable = unyielding

impending = to be near at hand

repugnance = the state of being opposed

formidable = exciting fear by size or strength

indeterminate = unclear or unable to be determined

orthodox = following established traditions and beliefs

superficial = concerned only with what is not necessarily real

remonstrance = statement of reasons against an idea or something

- ✓ Buck had a vague feeling of impending doom.
- ✓ Our vacation plan is still at an indeterminate stage.
- ✓ The crime scene aroused repugnance and aversion in him.
- ✓ Her criticism was perfectly innocuous, but the man was still offended.
- ✓ The teacher preferred a more orthodox approach to the problem.
- ✓ His article addressed the issue on a superficial level.
- ✓ Jerry's formidable opponent was nearly six feet tall.
- ✓ Without any remonstrance or objection, the court granted the matter.
- ✓ Population growth was inexorable as medical technology improved.
- ✓ Her wide gold-colored belt was the most salient thing about her outfit.

The Call of the Wild III Matching

Lesson 55

Can you match the words with their definitions?

impending indeterminate repugnance
remonstrance salient superficial orthodox
innocuous INEXORABLE formidable

______________ = the state of being opposed

______________ = unclear or unable to be determined

______________ = following established traditions and beliefs

______________ = exciting fear by size or strength

______________ = to be near at hand

______________ = conspicuous

______________ = unyielding

______________ = harmless

______________ = concerned only with what is not necessarily real

______________ = statement of reasons against an idea or something

The Call of the Wild III Spelling

Lesson 56

statement of reasons against an idea or something

R						T	R			C	E

the state of being opposed

		P	U		N				

concerned only with what is not necessarily real

				R			C			L

exciting fear by size or strength

F				I					E

unyielding

			X	O			B		

to be near at hand

I		P			D			

following established traditions and beliefs

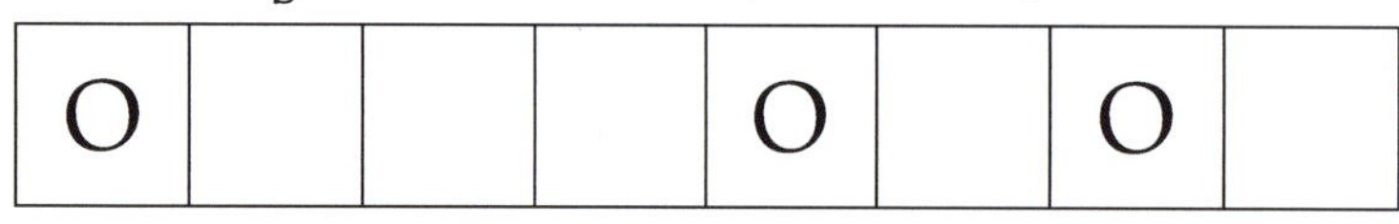

The Call of the Wild IV Vocabulary

Lesson 57

Review and read aloud the words and their meanings.

pertinacity = persistency

peremptorily = arrogantly

palpitant = trembling

plethoric = excessively full

feigned = pretended; faked

commingled = mixed together

expediency = convenience; a purposeful, focused hurry

tangible = real; able to be touched

- ✓ The old teddy bear was a tangible comfort to the child.
- ✓ The book contained fact commingled with fiction.
- ✓ Nowadays television is plethoric with reality shows.
- ✓ He was indignant that the director peremptorily rejected his proposal.
- ✓ With a feigned look of unconcern on her face, she tried to behave as normal as possible.
- ✓ Some companies see to the welfare of their employees from a sense of duty and business expediency.
- ✓ In the end Buck's pertinacity was rewarded; for the wolf, finding that no harm was intended, finally sniffed noses with him.
- ✓ As the moose were coming into the land, other kinds of life were coming in. Forest and stream and air seemed palpitant with their presence.

The Call of the Wild IV Matching

Lesson 58

Can you match the words with their definitions?

peremptorily	trembling
plethoric	pretended; faked
feigned	real; able to be touched
pertinacity	persistency
expediency	arrogantly
palpitant	mixed together
tangible	excessively full
commingled	convenience; a purposeful, focused hurry

Fun trivia for you! True or False?

The main character is a dog named Buck.

Buck ran away from his home in Alaska.

In Alaska, Buck was trained to be a sled dog.

Buck began to hate Spitz after Curly's death.

Level 5 Vocabulary Review I

Lesson 62

Choose the word from its definition. (The solution is on page 102.)

very great

○ tangible ○ immense ○ salable ○ unfettered

absolutely necessary

○ reproach ○ vigilant ○ amply ○ indispensable

deserving of honor and respect

○ venerable ○ venture ○ insinuate ○ languid

to make an opposite statement

○ aquatic ○ contradict ○ capsize ○ precocious

contemptuous ridicule or mockery

○ derision ○ clamorous ○ coax ○ array

a strong desire to do or to achieve something

○ sultry ○ mock ○ ingratitude ○ ambition

to be silent because you are in a bad mood, to pout

○ sulk ○ perversity ○ cultivate ○ abate

feeling or showing anger because of an injustice

○ somber ○ indignant ○ generation ○ inexplicable

Level 6 Vocabulary Review I

Lesson 62

Choose the word from the definition or the definition from the word. (The solution is on page 102.)

make or become wider, larger, or more open

○ vaguely ○ dilating ○ interrogative ○ ingratitude

the path of a flying object

○ trajectory ○ fathom ○ mock ○ transcend

degrade, profane or desecrate

○ menagerie ○ contradict ○ profanation ○ subdue

placed or inserted between two things

○ interposed ○ teeming ○ detriment ○ venerable

to come to know, have a fair knowledge of

○ miscellany ○ acquainted ○ impetuous ○ piteous

an adult male singing voice between tenor and bass

○ verdict ○ capsize ○ salable ○ baritone

tumultuous

○ feeling or showing anger because of an injustice

○ making a loud, confused noise; uproarious

○ ungrateful, not showing thanks for a kindness done to you

Level 5 Vocabulary Review II

Lesson 63

Choose the word from its definition. (The solution is on page 102.)

not clearly or precisely

○ insinuate ○ derision ○ vaguely ○ venerable

slow, lacking in interest

○ languid ○ transcend ○ contradict ○ indispensable

pathetic or deserving pity

○ capsize ○ piteous ○ simile ○ tangible

a cause of harm or damage

○ unfettered ○ detriment ○ aghast ○ generation

the state of being corrupted

○ inexplicable ○ insinuate ○ perversity ○ amply

impossible to repress, control

○ irrepressible ○ verdict ○ menagerie ○ unavailing

save money by doing things in a way that costs less

○ venerable ○ derision ○ apathy ○ economize

try to acquire or develop (a quality, sentiment, or skill)

○ interrogative ○ indignant ○ mock ○ cultivate

The Spy I Vocabulary

Lesson 67

Review and read aloud the words and their meanings.

petulance = irritability

evince = show, exhibit

ardently = fervently, passionately

unerring = always right or accurate

soliloquizing = speaking by oneself

contemptuous = scornful, expressing disdain

eloquence = fluent or persuasive speech or writing

portentous = done in a pompous or overly solemn manner

capricious = given to sudden changes in mood or behavior

sublime = of such excellence as to inspire great admiration

- ✓ This author's portentous, self-congratulating tone is annoying.
- ✓ The art display in the university museum is positively sublime.
- ✓ Your petulance made everyone in the room uncomfortable.
- ✓ The chairman is admired for his intelligence and eloquence.
- ✓ Captain Wallace is unerring in his judgment of the seas.
- ✓ The actor in the lead role soliloquizes often and well.
- ✓ Mabel looked at Hank with contemptuous eyes.
- ✓ His face evinced confidence and composure.
- ✓ Elaine studied ardently to pass the exam.
- ✓ Nero was a cruel and capricious emperor.

The Spy I Matching

Match the words with their definitions.

soliloquizing **sublime** EVINCE
portentous eloquence
capricious ARDENTLY petulance
contemptuous unerring

JAMES F. COOPER

_______________ = show, exhibit

_______________ = irritability

_______________ = speaking by oneself

_______________ = fervently, passionately

_______________ = always right or accurate

_______________ = scornful, expressing disdain

_______________ = fluent or persuasive speech or writing

_______________ = done in a pompous or overly solemn manner

_______________ = given to sudden changes in mood or behavior

_______________ = of such excellence as to inspire great admiration

The Spy I Crossword

Lesson 80

Across

5. show, exhibit
6. fervently, passionately
8. done in a pompous or overly solemn manner
9. of such excellence as to inspire great admiration

Down

1. irritability
2. always right or accurate
3. scornful, expressing disdain
4. fluent or persuasive speech or writing
7. given to sudden changes in mood or behavior

The Spy II Vocabulary

Lesson 81

Review and read aloud the words and their meanings.

relinquishing = voluntarily ceasing to keep or claim; giving up

conveyance = a means of transportation

impelled = to force or urge to do something

reprobate = an unprincipled person, scoundrel

prudence = caution with regard to practical matters

defrauded = illegally obtain money through deception

ostentatiously = with a boastful showiness, conspicuously

wafted = to carry lightly through air or over water

destitute = without the basic necessities of life

- ✓ Esmerelda is <u>relinquishing</u> control of her family's business next week.
- ✓ One tiny clue <u>impelled</u> Detective Alexander to investigate further.
- ✓ <u>Prudence</u> in financial matters is a key requirement for this job.
- ✓ Barges are used for the <u>conveyance</u> of coal along the Ohio River.
- ✓ When he died, his family was left completely <u>destitute</u>.
- ✓ Shirley was falsely accused of <u>defrauding</u> her clients.
- ✓ The aroma of bread in the oven <u>wafted</u> upstairs.
- ✓ I dressed <u>ostentatiously</u> when I was young.
- ✓ A <u>reprobate</u> is not to be trusted.

The Spy II Matching

Lesson 91

Match the words with their definitions.

A prudence	**B** impelled	**C** wafted
D destitute	**E** conveyance	**F** defrauded
G reprobate	**H** ostentatiously	**I** relinquishing

______ voluntarily ceasing to keep or claim; giving up

______ a means of transportation

______ to force or urge to do something

______ an unprincipled person, scoundrel

______ caution with regard to practical matters

______ illegally obtain money through deception

______ with a boastful showiness, conspicuously

______ to carry lightly through air or over water

______ without the basic necessities of life

All greatness of character is dependent on individuality. The man who has no other existence than that which he partakes in common with all around him, will never have any other than an existence of mediocrity.

– James F. Cooper

The Spy II Crossword

Lesson 96

Across

1. voluntarily ceasing to keep or claim; giving up
4. caution with regard to practical matters
5. illegally obtain money through deception
6. to carry lightly through air or over water
7. to force or urge to do something

Down

1. an unprincipled person, scoundrel
2. a means of transportation
3. without the basic necessities of life

Analogies Matching I

Lesson 111

Analogies are word comparisons. Choose the correct word to complete each relationship. (The solution is on page 102.)

toe : foot :: finger : ________________

○ hand ○ head ○ face

eyes : see :: ________________ : chew

○ feet ○ hands ○ teeth

stem : flower :: ________________ : tree

○ branch ○ trunk ○ seed

fish : gills :: human : ________________

○ hair ○ stomach ○ lungs

cold : ice :: heat : ________________

○ fire ○ water ○ wood

pig : pork :: cow : ________________

○ beef ○ dairy ○ fruit

Analogies Matching II

Lesson 113

Analogies are word comparisons. Choose the correct word to complete each relationship. (The solution is on page 102.)

pencil : write :: knife : ________________

○ wash ○ cut ○ bake

hungry : eat :: tired : ________________

○ clean ○ jog ○ rest

glasses : see :: cane : ________________

○ sleep ○ think ○ walk

boring : interesting :: rare : ________________

○ occasional ○ unusual ○ common

key : unlock :: scale : ________________

○ wear ○ weigh ○ think

cub : bear :: kitten : ________________

○ cat ○ dog ○ mouse

Analogies Matching III

Lesson 116

Analogies are word comparisons. Choose the pair of words whose relationship is similar to the given pair. (The solution is on page 102.)

match : fire

○ key : car ○ money : wallet ○ bread : bakery

candy : sweet

○ water : drink ○ dog : fast ○ fire : hot

teacher : educate

○ sky : blue ○ car : drive ○ doctor : heal

whale : ocean

○ car : garage ○ lion : jungle ○ chair : table

television : watch

○ book : smile ○ novel : read ○ notebook : write

brother : sibling

○ sister : son ○ doctor : nurse ○ father : parent

Treasure Island Trivia I

Lesson 120

Can you answer the following questions? (The solution is on page 103.)

The narrator of the story is

○ Dr. Livesey ○ Jim Hawkins ○ Long John Silver

The story begins in

○ Treasure Island ○ The Admiral Benbow Inn

Captain Flint's first name is

○ William ○ James ○ John

Jim's father died of

○ drinking ○ illness ○ cancer

The Dead Man's Chest is

○ an island ○ a chest ○ a ship

The treasure was initially buried on

○ Spy-glass Hill ○ Skeleton Hill ○ Black Hill Cove

The gold buried on Treasure Island originally came from

○ Port Bello ○ Savannah ○ Madrid

Treasure Island was written by

○ J.K. Rolling ○ Jonathan Swift ○ Robert L. Stevenson

Treasure Island Trivia II

Lesson 120

Can you answer the following questions? (The solution is on page 103.)

Flint died in

○ Havana ○ Savannah ○ Madrid

Long John Silver's nickname is

○ Silver ○ Captain ○ Barbecue

The name of the ship is

○ The Sea Chest ○ The Hispanolia ○ The Admiral

Billy Bones worked for

○ Dr. Livesey ○ Captain Flint ○ Long John Silver

Long John Silver's occupation is

○ a captain ○ a cook ○ a doctor

The first pirate that Jim encounters is

○ Long John Silver ○ Billy Bones ○ Israel Hans

The flag that represents the pirates is the

○ Evil Skeleton ○ Jolly Roger ○ French Flag

Long John's parrot repeats the phrase

○ Pieces of four ○ Pieces of six ○ Pieces of eight

The Spy II Fill-in-the-Blanks

Lesson 121

Fill in the blanks to complete the sentences. Change the word forms if necessary.

ostentatiously	**RELINQUISH**	reprobate	prudence	
defraud	**conveyance**	impelled	**wafted**	*destitute*

1. ______________ in financial matters is a key requirement for this job.
2. Esmerelda is ____________ control of her family's business next week.
3. Barges are used for the ____________ of coal along the Ohio River.
4. When he died, his family was left completely ______________.
5. One tiny clue ______________ Detective Alexander to investigate further.
6. Shirley was falsely accused of ______________ her clients.
7. The aroma of bread in the oven ______________ upstairs.
8. I dressed ______________ when I was young.
9. A ______________ is not to be trusted.

Penrod and Sam II Matching

Lesson 122

Can you match the words with their definitions?

A ferocity	**B** plaintively	**C** mollify
D truculence	**E** devoid	**F** converged
G grievance	**H** stratagem	**I** dexterity
J precociously	**K** unobtrusively	

_______ a plan or scheme

_______ entirely lacking or free from

_______ inconspicuous, unassertive

_______ appease the anger or anxiety of (someone)

_______ obstreperous and defiant aggressiveness

_______ skill in performing tasks, especially with the hands

_______ expressing sorrow or melancholy; mournful

_______ the state or quality of being ferocious; fierceness

_______ come together from different directions so as eventually to meet

_______ unusually advanced or mature in development, especially mental development

_______ a real or imagined wrong or other cause for complaint or protest, especially unfair treatment

Level 6 Vocabulary Review II

Lesson 123

Choose the word from the definition. (The solution is on page 103.)

artificial light; someone who inspires

○ exasperated ○ demolish ○ luminary ○ slumber

marked or dotted with color

○ smug ○ flecked ○ emulate ○ tirelessly

eagerness, interest in

○ keen ○ impudent ○ admire ○ scorn

overcome, quieten, or bring under control

○ gleam ○ disrepair ○ intently ○ subdue

understand after much thought

○ luminary ○ imposing ○ fathom ○ exasperated

a group or collection of different items

○ loiter ○ miscellany ○ deliberation ○ pungent

consisting of small parts that are disconnected or incomplete

○ fragmentary ○ intimidate ○ contempt ○ vigorous

knobby, rough, and twisted, especially with age

○ atrocious ○ scold ○ gnarled ○ eminent

The Talisman Vocabulary

Lesson 131

Review and read aloud the words and their meanings.

sinewy = tough, firm, braided, or resilient

laurels = honor; the state of being honored

mortification = a feeling of humiliation or shame

sordid = morally ignoble or base; vile; dirty or filthy

mitigation = the process of becoming milder, gentler, or less severe

taciturn = reserved or uncommunicative in speech; saying little

audacious = extremely bold or daring; fearless; extremely original; recklessly bold in defiance of convention

- ✓ He is reserved and taciturn by nature.
- ✓ Their next move was even more audacious and bold.
- ✓ Millie's team enjoyed the laurel of victory once again.
- ✓ Mitigation of injury risk is the job of our safety officer.
- ✓ With his large frame and sinewy arms, Billy is a great fighter.
- ✓ The movie described the sordid history of the mafia since 1950.
- ✓ If I could have explained these things to Miss Caroline, I would have saved myself some inconvenience and Miss Caroline's subsequent mortification, but it was beyond my ability to explain things as well as Atticus, so I said, "You are shamin' him, Miss Caroline."

  \- Harper Lee, *To Kill a Mockingbird*

Fun trivia for you! True or False?

The Talisman takes place at the end of the Third Crusade.

The Talisman Matching

Lesson 136

Can you match the words with their definitions?

sordid   taciturn   laurels   MITIGATION
sinewy   MORTIFICATION   audacious

_______________ = tough, firm, braided, or resilient

_______________ = honor; the state of being honored

_______________ = a feeling of humiliation or shame

_______________ = morally ignoble or base; vile; dirty or filthy

_______________ = the process of becoming, milder, gentler, or less severe

_______________ = reserved or uncommunicative in speech; saying little

_______________ = extremely bold or daring; extremely original; fearless; recklessly bold in defiance of convention

Fun trivia for you! True or False?

The main character in the book is Sir Kenneth.

Sir Kenneth came from England to fight in the Crusade.

The Talisman Multiple Choice

Lesson 138

Choose the word from the definition.

extremely bold or darling; fearless; extremely original

○ salable ○ furrows ○ audacious ○ inverted

reserved or uncommunicative in speech; saying little

○ taciturn ○ tangible ○ perpetually ○ capsize

the process of becoming milder, gentler, or less severe

○ mitigation ○ amply ○ cultivate ○ piteous

morally ignoble or base; vile; dirty or filthy

○ venture ○ precocious ○ verdict ○ sordid

a feeling of humiliation or shame

○ grievance ○ disconsolate ○ mortification ○ dispersed

honor; the state of being honored

○ laurels ○ verdict ○ generation ○ insinuate

tough, firm, braided, or resilient

○ amiable ○ sinewy ○ ambition ○ clamorous

For success, attitude is equally as important as ability.

– Sir Walter Scott

Penrod and Sam I Spelling

primitive; savagely cruel; exceedingly brutal

B		R					C

sounding sad and mournful

P				N			V	

capable of producing a deep or ringing sound

S		N					S

having a harsh, grating sound

G	U			U			L

resounding or echoing

R		S					T

offensively bold; impertinent

I	M			D		N	

depressing thoughts

B	R						G	S

PENROD AND SAM
BOOTH TARKINGTON

PUBLISHED IN 1916

Penrod and Sam I Crossword

THE PREQUEL PUBLISHED IN 1914

Across

4. resounding or echoing
7. offensively bold; impertinent

Down

1. depressing thoughts
2. having a harsh, grating sound
3. sounding sad and mournful
5. savagely cruel; exceedingly brutal
6. capable of producing a deep or ringing sound

Penrod and Sam I Word Jumble

Unscramble the jumbled words.

depressing thoughts

DIOORNBSG → ________________________

sounding sad and mournful

VLIATNPEI → ________________________

savagely cruel; exceedingly brutal

CRARIBBA → ________________________

having a harsh, grating sound

TTGUURAL → ________________________

offensively bold; impertinent

NIDMPUET → ________________________

resounding or echoing

ORNSTENA → ________________________

capable of producing a deep or ringing sound

SOORSUNO → ________________________

Nothing ever becomes real till it is experienced. - John Keats

Penrod and Sam I Word Search

Find the hidden words and explain their meanings. The words can go in any direction, even backwards! (The solution is on page 103.)

S	J	R	A	E	M	R	D	Q	V
X	G	H	E	R	C	T	C	G	E
G	M	N	I	H	Y	V	H	L	V
S	U	T	I	U	K	I	J	B	I
O	F	T	Q	D	P	C	A	A	T
N	Q	L	T	D	O	R	H	T	N
O	V	A	H	U	B	O	N	L	I
R	O	O	U	A	R	E	R	T	A
O	N	G	R	G	D	A	D	B	L
U	N	I	Z	U	A	E	L	L	P
S	C	M	P	Z	I	J	U	Z	N
C	W	M	J	X	U	L	W	V	Z
G	I	R	E	S	O	N	A	N	T
N	C	V	G	I	N	J	L	I	R
L	T	B	Y	O	Z	Z	S	B	K

barbaric

impudent

sonorous

broodings

plaintive

guttural

resonant

Penrod and Sam I Fill-in-the-Blanks

Fill in the blanks to complete the sentences. Change the word forms if necessary.

broodings   **barbaric**   resonant

impudent   **GUTTURAL**   sonorous   plaintive

These ________________ helped a little; but it was a severe morning, and on his way home at noon he did not recover heart enough to practice the bullfrog's croak, the craft that Sam Williams had lately mastered to inspiring perfection. This ________________ accomplishment Penrod had determined to make his own. At once ________________ and ________________, ________________ yet ________________, with a ________________ twang like the plucked string of a Congo war-fiddle, the sound had fascinated him. It is made in the throat by processes utterly impossible to describe in human words, and no alphabet as yet produced by civilized man affords the symbols to vocalize it to the ear of imagination. "Gunk" is the poor makeshift that must be employed to indicate it.

Hope is the thing with feathers that perches in the soul -- and sings the tunes without the words -- and never stops at all. - *Emily Dickinson*

Penrod and Sam I Multiple Choice

Choose the word from the definition.

depressing thoughts

○ impudent ○ furrow ○ broodings ○ coax

offensively bold; impertinent

○ fathom ○ impudent ○ sulk ○ venture

capable of producing a deep or ringing sound

○ keen ○ sonorous ○ aquatic ○ piteous

primitive; savagely cruel; exceedingly brutal

○ subdue ○ populous ○ barbaric ○ capsize

sounding sad and mournful

○ agitated ○ plaintive ○ sordid ○ reproach

resounding or echoing

○ rued ○ coarse ○ assure ○ resonant

having a harsh, grating sound

○ sonorous ○ guttural ○ grove ○ torment

To strive, to seek, to find, and not to yield. - Alfred Lord Tennyson

Penrod and Sam II Spelling

entire lacking or free from

D		V			

appease the anger or anxiety of

	O			I		Y

the state or quality of being ferocious

F			O	C			

skill in performing tasks, especially with the hands

D		X					T	

a plan or scheme

S		R		T		G		

obstreperous and defiant aggressiveness

T			C				N	C	

expressing sorrow or melancholy; mournful

	L	A					V			Y

Penrod and Sam II Crossword

Across

4. the state or quality of being ferocious; fierceness
5. obstreperous and defiant aggressiveness
6. inconspicuous, unassertive
7. a plan or scheme

Down

1. entirely lacking or free from
2. skill in performing tasks, especially with the hands
3. a real or imagined wrong or other cause for complaint or protest, especially unfair treatment

Penrod and Sam II Word Jumble

Unscramble the jumbled words.

come together from different directions so as eventually to meet

ONERGCEVD → ____________________

skill in performing tasks, especially with the hands

DXEIRYETT → ____________________

the state or quality of being ferocious; fierceness

ECOITFYR → ____________________

appease the anger or anxiety of (someone)

FYLOMIL → ____________________

expressing sorrow or melancholy; mournful

NTLIPVYILAE → ____________________

entirely lacking or free from

EDVIOD → ____________________

inconspicuous, unassertive

SYNIEVUURLBOT → ____________________

a plan or scheme

EGATTRMSA → ____________________

Penrod and Sam II Word Search

Find the hidden words and explain their meanings. The words can go in any direction, even backwards! (The solution is on page 103.)

E	F	E	R	O	C	I	T	Y	P
C	K	C	D	K	O	M	E	L	L
N	X	N	C	P	N	U	N	S	A
E	H	A	E	Z	V	N	S	U	I
L	Y	V	M	Q	E	O	T	O	N
U	T	E	O	C	R	B	R	I	T
C	I	I	L	K	G	T	A	C	I
U	R	R	L	I	E	R	T	O	V
R	E	G	I	L	D	U	A	C	E
T	T	F	F	R	D	S	G	E	L
F	X	C	Y	E	V	I	E	R	Y
L	E	E	V	O	K	V	M	P	U
E	D	O	X	N	J	E	C	J	X
B	I	M	Y	F	K	L	V	C	N
D	O	S	W	F	K	Y	K	D	L
D	X	Y	U	G	Z	K	E	L	M

unobtrusively
stratagem
devoid
mollify
truculence
dexterity
plaintively
ferocity
converged
precociously
grievance

Penrod and Sam II Fill-in-the-Blanks

Fill in the blanks to complete the sentences. Change the word forms if necessary.

converged STRATAGEM unobtrusively
mollify ferocity devoid grievance
dexterity plaintively precocious

1. Ambush is an effective military ________________.

2. She's always singing some old song ________________.

3. The tiger attacked the antelope with ________________.

4. Eric's speech was ________________ of warmth and feeling.

5. There was a signpost where the two paths ________________.

6. You need manual ________________ to be good at video games.

7. Will this be enough to ________________ the outraged citizens?

8. He spoke with the king quietly and ________________ for a while.

9. From an early age she displayed a ________________ talent for music.

10. Does your organization have a formal ________________ procedure?

SO LONG AS WE CAN LOSE ANY HAPPINESS, WE POSSESS SOME.

– BOOTH TARKINGTON

Penrod and Sam II Multiple Choice

Choose the word from the definition.

expressing sorrow or melancholy; mournful

○ impudent ○ plaintively ○ broodings ○ barbaric

a plan or scheme

○ devoid ○ converged ○ resonant ○ stratagem

entirely lacking or free from

○ diffuse ○ miscellany ○ devoid ○ acquainted

inconspicuous, unassertive

○ fathom ○ unobtrusively ○ subdue ○ guttural

skill in performing tasks, especially with the hands

○ grievance ○ dexterity ○ ferocity ○ trajectory

come together from different directions so as eventually to meet

○ rued ○ coarse ○ converged ○ dilating

appease the anger or anxiety of (someone)

○ mollify ○ interposed ○ peck ○ russet

obstreperous and defiant aggressiveness

○ sanctify ○ effluvia ○ grove ○ truculence

The Call of the Wild I Spelling

to intertwine

I						A		E

one's children

	R		G			Y

to cut, especially by tearing

L				R	A		

to come together and be friends

C			C					T	

character trait of someone who loves to talk about himself

		O			S	M

controlling; urgent

I				R				S

the legal transfer of property

C			V		Y			C	

The Call of the Wild I Word Jumble

Unscramble the jumbled words.

one's children

OYPNGRE → ____________________

to intertwine

ICTNELAER → ____________________

to spread everywhere

VPEEDRA → ____________________

controlling; urgent

ORIPMEUIS → ____________________

unclear; to hide from view by covering

SCBUOER → ____________________

character trait of someone who loves to talk about himself

MOEISGT → ____________________

to cut, especially by tearing

EAECALRT → ____________________

to come together and be friends

ILICTAOCNE → ____________________

The Call of the Wild I Fill-in-the-Blanks

Fill in the blanks to complete the sentences. Change the word forms if necessary.

progeny	egotism	pervade	
interlace	imperious	conciliate	obscure
CONVEYANCE	latent	lacerate	

1. ________________ of land is taxed in most states.

2. Red and blue ribbons were ________________ with one another.

3. Some viruses can remain ________________ in the body for years.

4. Her hand was ________________ by a broken glass bottle.

5. The family lived ________________ in the shadow of the forest.

6. His ________________ and arrogance annoyed everyone around him.

7. An atmosphere of excitement ________________ the town.

8. The king was notorious for being ________________ and having an explosive temper.

9. The farmer worked hard and saved money so his ________________ might have an easier life.

10. After a long argument, Sally ________________ her brother by offering him some candy.

The Call of the Wild I Multiple Choice

Choose the word from the definition.

to intertwine

○ converge ○ interlace ○ interpose ○ fathom

character trait of someone who loves to talk about himself

○ ferocity ○ interrogative ○ menagerie ○ egotism

controlling; urgent

○ sonorous ○ imperious ○ tumultuous ○ impudent

unclear; to hide from view by covering

○ devoid ○ venerable ○ obscure ○ teeming

one's children

○ broodings ○ progeny ○ generation ○ grove

to come together and be friends

○ reverberate ○ subdue ○ conciliate ○ sanctify

to spread everywhere

○ pervade ○ abate ○ coax ○ adorn

to cut, especially by tearing

○ dilate ○ diffuse ○ dispute ○ lacerate

The Call of the Wild II Spelling

character trait of revealing private information

I		D	I					T	I		

delicate or refined; hard to please in matters of taste

			T		D			U	S

aggressive

B		L	L				R			

serving in the place of someone or something else

V		C			I			

to pretend injury or sickness to avoid responsibility

			I	N			R

rhythmic movement or series of sounds

C					C	E

a messenger

C			R			R

The Call of the Wild II Word Jumble

Unscramble the jumbled words.

a messenger

IEURORC → ____________________

a reversal in development from a higher to lower state

SREOIOSNREGTR → ____________________

aggressive

LTNEEIRBGEL → ____________________

character trait of revealing private information

INTOCSEIDIRN → ____________________

delicate or refined; hard to please in matters of taste

IDOTIUFSSA → ____________________

difficult

RAUSUDO → ____________________

rhythmic movement or series of sounds

NEAEDCC → ____________________

serving in the place of someone or something else

SURAOICIV → ____________________

The Call of the Wild II Fill-in-the-Blanks

Fill in the blanks to complete the sentences. Change the word forms if necessary.

cadence	belligerent	**malinger**	
fastidious	courier	*arduous*	**indiscretion**
retrogression	**CONSTERNATION**	*vicarious*	

1. The chef is very ______________ about the food he makes.

2. He gets a ______________ thrill out of watching adventure movies.

3. ______________ to a child-like state is possible with some illnesses.

4. She tried to calm him down but he remained ______________.

5. The informant did not expect the detective's ______________.

6. During the game, the cheerleaders danced in a rapid ______________.

7. The CEO's shocking decision caused ______________ in the company.

8. The doctor accused Tommy of ______________, but he was really sick.

9. After a long and ______________ day's work, he was completely exhausted.

10. A ______________ brought the soldiers' letters from the battleground to the city.

The Call of the Wild II Multiple Choice

Choose the word from the definition.

serving in the place of someone or something else

○ vicarious ○ guttural ○ sultry ○ vigilant

to pretend injury or sickness to avoid responsibility

○ peck ○ economize ○ malinger ○ capsize

sudden confusion or amazement

○ languid ○ consternation ○ dismay ○ profanation

delicate or refined; hard to please in matters of taste

○ fastidious ○ acquainted ○ disconsolate ○ inverted

a reversal in development from a higher to lower state

○ dexterity ○ perversity ○ ignorance ○ retrogression

rhythmic movement or series of sounds

○ insinuation ○ cadence ○ resonant ○ russet

difficult

○ arduous ○ plaintive ○ grievance ○ precocious

character trait of revealing private information

○ indiscretion ○ stratagem ○ luminary ○ magnanimous

The Call of the Wild III Word Jumble

Unscramble the jumbled words.

statement of reasons against an idea or something

MOAREENCNTSR → ____________________

to be near at hand

GNIDENMIP → ____________________

exciting fear by size or strength

ABODERMLIF → ____________________

unyielding

LNEOXIEABR → ____________________

the state of being opposed

RCNEEAPUGN → ____________________

harmless

OIONSCUNU → ____________________

unclear or unable to be determined

NEMEIINDETATR → ____________________

following established traditions and beliefs

RHXODTOO → ____________________

The Call of the Wild III Fill-in-the-Blanks

Fill in the blanks to complete the sentences. Change the word forms if necessary.

impending	indeterminate	repugnance	
remonstrance	salient	*superficial*	orthodox
innocuous	INEXORABLE	formidable	

1. Our vacation plan is still at an ______________ stage.

2. Buck had a vague feeling of ______________ doom.

3. His article addressed the issue on a ______________ level.

4 The crime scene aroused ______________ and aversion in him.

5. The teacher preferred a more ___________ approach to the problem.

6. Jerry's ______________ opponent was nearly six feet tall.

7. Her criticism was perfectly ______________ but the man was still offended.

8. Without any ______________ or objection, the court granted the matter.

9. Population growth was ______________ as medical technology improved.

10. Her wide gold-colored belt was the most ______________ thing about her outfit.

The Call of the Wild III Multiple Choice

Choose the word from the definition.

the state of being opposed

○ trajectory ○ mock ○ repugnance ○ malinger

statement of reasons against an idea or something

○ simile ○ foliage ○ miscellany ○ remonstrance

exciting fear by size or strength

○ cumulative ○ formidable ○ tangible ○ truculence

harmless

○ innocuous ○ keen ○ somber ○ aghast

following established traditions and beliefs

○ amply ○ orthodox ○ barbaric ○ populous

unclear or unable to be determined

○ dispersed ○ clamorous ○ gnarled ○ indeterminate

conspicuous

○ languid ○ salient ○ indispensable ○ immense

concerned only with what is not necessarily real

○ superficial ○ piteous ○ recumbent ○ unavailing

The Call of the Wild IV Spelling

mixed together

C					N				D

convenience; a purposeful, focused hurry

	X			D					Y

arrogantly

		R	O					R			Y

persistency

P			T			A				

trembling

P			P					T

real; able to be touched

			G			L	E

pretended; faked

F					E	D

The Call of the Wild IV Word Jumble

Unscramble the jumbled words.

mixed together

IMOGCEMNDL → ____________________

real; able to be touched

ITNAGBEL → ____________________

convenience; a purposeful, focused hurry

NEIPDXCEYE → ____________________

pretended; faked

IDNEGFE → ____________________

the act of paying careful attention to details

MANSTTOSIIRIN → ____________________

excessively full

ORHPELICT → ____________________

persistency

TEARTNCIIYP → ____________________

arrogantly

RPYMILOTEPER → ____________________

The Call of the Wild IV Fill-in-the-Blanks

Fill in the blanks to complete the sentences. Change the word forms if necessary.

peremptorily	plethoric	*feigned*	pertinacity
expediency	**PALPITANT**	**tangible**	*commingled*

1. The old teddy bear was a ________________ comfort to the child.

2. The book contained fact ________________ with fiction.

3. Nowadays television is ________________ with reality shows.

4. He was indignant that the director ________________ rejected his proposal.

5. With a ________________ look of unconcern on her face, she tried to behave as normal as possible.

6. Some companies see to the welfare of their employees from a sense of duty and business ________________.

7. In the end Buck's ________________ was rewarded; for the wolf, finding that no harm was intended, finally sniffed noses with him.

8. As the moose were coming into the land, other kinds of life were coming in. Forest and stream and air seemed ________________ with their presence.

The Call of the Wild IV Multiple Choice

Choose the word from the definition.

persistency

- ○ derision
- ○ baritone
- ○ verdict
- ○ pertinacity

arrogantly

- ○ peremptorily
- ○ perpetually
- ○ maliciously
- ○ unobtrusively

pretended; faked

- ○ impetuous
- ○ aquatic
- ○ feigned
- ○ agitated

convenience; a purposeful, focused hurry

- ○ contradiction
- ○ venture
- ○ expediency
- ○ ambition

mixed together

- ○ unfettered
- ○ commingled
- ○ cultivated
- ○ sulked

the act of paying careful attention to details

- ○ detriment
- ○ hesitation
- ○ ministrations
- ○ knoll

trembling

- ○ mollify
- ○ transcend
- ○ palpitant
- ○ subside

excessively full

- ○ plethoric
- ○ rued
- ○ flecked
- ○ fragmentary

The Spy I Spelling

always right or accurate

U	N					N	G

of such excellence as to inspire great admiration

S			L			E

fervently, passionately

A	R					L	Y

fluent or persuasive speech or writing

E			Q			N		E

given to sudden changes in mood or behavior

C	A	P	R	I	C	I	O	U	S

done in a pompous or overly solemn manner

P		R	T			T			

scornful, expressing disdain

C			T				T	U			

The Spy I Word Jumble

Unscramble the jumbled words.

irritability

NELACTUPE → ______________________

show, exhibit

EIVNCE → ______________________

speaking by oneself

SOONIQGLZUILI → ______________________

fervently, passionately

READTLYN → ______________________

always right or accurate

ERNURNIG → ______________________

scornful, expressing disdain

MCOONSUPTETU → ______________________

fluent or persuasive speech or writing

EUQNCEOEL → ______________________

done in a pompous or overly solemn manner

OOERNSPTTU → ______________________

The Spy I Fill-in-the-Blanks

Fill in the blanks to complete the sentences. Change the word forms if necessary.

evince	PORTENTOUS	soliloquize	*sublime*	**ardently**
capricious	contemptuous	**petulance**	eloquence	**UNERRING**

1. This author's _______________, self-congratulating tone is annoying.

2. Your _______________ made everyone in the room uncomfortable.

3. The art display in the university museum is positively _______________.

4. The chairman is admired for his intelligence and _______________.

5. Captain Wallace is _______________ in his judgment of the seas.

6. The actor in the lead role _______________ often and well.

7. His face _______________ confidence and composure.

8. Mabel looked at Hank with _______________ eyes.

9. Elaine studied _______________ to pass the exam.

10. Nero was a cruel and _______________ emperor.

IGNORANCE AND SUPERSTITION EVER BEAR A CLOSE AND MATHEMATICAL RELATION TO EACH OTHER. – **JAMES F. COOPER**

The Spy II Spelling

to force or urge to do something

I		P					D

caution with regard to practical matters

P			D			C	E

without the basic necessities of life

D			T			U	T	

a means of transportation

			V			A		C	

an unprincipled person, scoundrel

R		P				A	T	

to carry lightly through air or over water

			T	E	

illegally obtain money through deception

D			R			D		D

The Spy II Word Jumble

Unscramble the jumbled words.

a means of transportation

EYVNCNAOEC → ____________________

an unprincipled person, scoundrel

APBORERET → ____________________

caution with regard to practical matters

EECNDPRU → ____________________

voluntarily ceasing to keep or claim; giving up

NRSIGILIQUNHE → ____________________

illegally obtain money through deception

EEFURADDD → ____________________

without the basic necessities of life

UITSEDTET → ____________________

to carry lightly through air or over water

FATEDW → ____________________

to force or urge to do something

MLLPEEDI → ____________________

Treasure Island Characters

Match the characters with their descriptions. (The solution is on page 104.)

A. Black Dog
B. Captain Smollet
C. Long John Silver
D. Ben Gunn
E. Squire Trelawney
F. Billy Bones
G. Jim Hawkins
H. Dr. Livesey
I. Mr. Arrow
J. O'Brien
K. Blind Pew
L. Israel Hands

C	a one-legged ship's cook
G	the narrator of the story; a bright, courageous boy
B	the captain of the expedition's ship, the Hispaniola
I	the first mate of the Hispaniola
H	a local physician who treated Jim's father
L	a pirate shot by Jim after he tried to kill Jim with a knife
J	a pirate who wears a red cap and is killed by Israel Hands.
E	a local Bristol nobleman who finances the treasure hunt
D	a pirate marooned by Captain Flint on Treasure Island
K	a deformed pirate who delivers the Black Spot death notice
F	a pirate who steals the map and sings, "Fifteen men on a dead man's chest."
A	a pirate who discovers Bone's hiding place and is almost killed in a fight in the inn parlor

Treasure Island Word Search I

Find the hidden words and explain their meanings. The words can go in any direction, even backwards! (The solution is on page 104.)

C	E	N	Q	C	G	O	B	B	P	R	Y	U	K
A	R	C	O	B	O	E	O	I	N	N	Q	O	T
P	U	H	V	T	A	M	R	L	I	P	O	P	E
T	S	E	C	C	E	A	P	T	O	B	O	I	K
A	A	S	H	O	T	L	U	A	G	T	G	H	S
I	E	T	B	E	V	M	E	O	S	X	S	S	U
N	R	R	S	W	I	E	L	K	Q	S	G	I	M
S	T	E	R	O	H	S	A	I	S	T	G	U	P
C	A	N	N	O	N	S	L	P	A	R	R	O	T
H	T	Y	P	B	B	W	W	A	R	S	D	Z	T
T	C	N	V	N	F	K	V	U	N	O	U	M	P
T	S	A	O	C	A	P	E	F	E	D	Q	K	I

ashore	chest	logbook	pistol	captain
beach	coast	musket	sail	island
cannons	compass	mutiny	ship	pirates
cape	cove	parrot	skeleton	treasure

Treasure Island Word Search II

Find the hidden words and explain their meanings. The words can go in any direction, even backwards! (The solution is on page 104.)

E	W	E	E	W	R	D	I	J	G	M	Y	N	S
D	P	F	K	O	G	I	G	Z	Z	A	A	B	R
A	O	X	H	C	G	G	S	X	S	S	N	P	A
P	W	C	S	E	A	M	E	N	H	T	U	O	S
S	N	M	L	T	L	J	U	Q	V	E	Q	K	T
A	E	V	N	P	F	R	T	S	Y	H	C	G	D
M	Y	I	I	N	W	V	O	I	K	S	B	O	S
H	L	H	L	I	A	S	V	E	P	E	C	R	I
F	S	M	C	R	E	W	J	A	D	K	T	I	L
F	O	U	V	L	P	N	D	K	E	Y	K	C	V
O	O	P	R	S	U	E	M	X	P	C	N	J	E
X	N	R	F	B	S	E	T	B	C	O	O	K	R

anchor	key	south	crew	flag
brush	map	spade	flint	ship
clue	mast	seamen	dig	silver
dock	sail	musket	jack	cook

The Talisman Spelling

extremely bold or daring; fearless; extremely original

A			A	C			U	

reserved or uncommunicative in speech; saying little

T				T		R	

the process of becoming milder, gentler, or less severe

M		T				T			N

honor; the state of being honored

	A		R			S

morally ignoble or base; vile; dirty or filthy

S			D		D

tough, firm, braided, or resilient

S				W	

The Talisman Crossword

"He brandished the weapon." "'Thou errest.' said Hakim."

Across

1. the process of becoming milder, gentler, or less severe
6. extremely bold or daring; fearless; extremely original; recklessly bold in defiance of convention

Down

2. reserved or uncommunicative in speech; saying little
3. morally ignoble or base; vile; dirty or filthy
4. honor; the state of being honored
5. tough, firm, braided, or resilient

The Talisman Word Jumble

Unscramble the jumbled words.

a feeling of humiliation or shame:

IICOFNIOAMTTR → ______________________

honor; the state of being honored:

AERLSLU → ______________________

tough, firm, braided, or resilient:

NEYISW → ______________________

morally ignoble or base; vile; dirty or filthy:

SODRDI → ______________________

reserved or uncommunicative in speech; saying little:

IACNURTT → ______________________

the process of becoming milder, gentler, or less severe:

OTGTINMIIA → ______________________

extremely bold or daring; fearless; extremely original:

COAADUUIS → ______________________

He that climbs the tall tree has won right to the fruit, He that leaps the wide gulf should prevail in his suit. – Sir Walter Scott

The Talisman Fill-in-the-Blanks

Fill in the blanks to complete the sentences. Change the word forms if necessary.

laurel TACITURN sordid audacious
mitigation sinewy mortification

1. He is reserved and ________________ by nature.
2. Their next move was even more ______________ and bold.
3. Millie's team enjoyed the ________________ of victory once again.
4. With his large frame and ________________ arms, Billy is a great fighter.
5. The movie described the _______________ history of the mafia since 1950.
6. ________________ of injury risk is the job of our safety officer.
7. If I could have explained these things to Miss Caroline, I would have saved myself some inconvenience and Miss Caroline's subsequent ________________, but it was beyond my ability to explain things as well as Atticus, so I said, "You are shamin' him, Miss Caroline."

– Harper Lee, *To Kill a Mockingbird*

Vocabulary Review Matching I

Match the words with their definitions.

conciliate	impelled	indiscretion	audacious
mortification	orthodox	eloquence	grievance
capricious	remonstrance	devoid	ministrations

__________________ = entirely lacking or free from

__________________ = to come together and be friends

__________________ = following established traditions and beliefs

__________________ = character trait of revealing private information

__________________ = statement of reasons against an idea or something

__________________ = the act of paying careful attention to details

__________________ = given to sudden changes in mood or behavior

__________________ = to force or urge to do something

__________________ = a feeling of humiliation or shame

__________________ = fluent or persuasive speech or writing

__________________ = a real or imagined wrong or other cause for complaint or protest, especially unfair treatment

__________________ = extremely bold or daring; fearless; extremely original; reckless bold in defiance of convention

Vocabulary Review Matching II

Match the words with their definitions.

sordid	plaintive	conveyance	barbaric
indeterminate	lacerate	retrogression	dexterity
destitute	converged	obscure	contemptuous

______________ = a means of transportation

______________ = sounding sad and mournful

______________ = to cut, especially by tearing

______________ = scornful, expressing disdain

______________ = without the basic necessities of life

______________ = unclear or unable to be determined

______________ = unclear; to hide from view by covering

______________ = primitive; savagely cruel; exceedingly brutal

______________ = morally ignoble or base; vile; dirty or filthy

______________ = skill in performing tasks, especially with the hands

______________ = a reversal in development from a higher to lower state

______________ = come together from different directions so as eventually to meet

Vocabulary Review Matching III

Match the words with their definitions.

egotism	mitigation	consternation	laurels
malinger	sonorous	vicarious	formidable
wafted	sublime	commingled	defrauded

_________________ = mixed together

_________________ = exciting fear by size or strength

_________________ = to pretend sickness to avoid responsibility

_________________ = capable of producing a deep or ringing sound

_________________ = serving in the place of someone else

_________________ = sudden confusion or amazement

_________________ = to carry lightly through air or over water

_________________ = of such excellence as to inspire great admiration

_________________ = illegally obtain money through deception

_________________ = honor; the state of being honored

_________________ = character trait of someone who loves to talk about himself

_________________ = the process of becoming milder, gentler, or less severe

Vocabulary Review Matching IV

Can you match the words with their definitions?

A. impending	B. courier	C. feigned	D. soliloquizing
E. palpitant	F. evince	G. salient	H. portentous
I. unerring	J. broodings	K. innocuous	L. plethoric
M. latent	N. ardently	O. taciturn	P. reprobate
Q. tangible	R. relinquish	S. pervade	T. interlace

______ not shown

______ a messenger

______ harmless

______ trembling

______ show, exhibit

______ to be near at hand

______ conspicuous

______ to intertwine

______ pretended; faked

______ excessively full

______ speaking by oneself

______ real; able to be touched

______ fervently, passionately

______ always right or accurate

______ depressing thoughts

______ to spread everywhere

______ reserved or uncommunicative in speech; saying little

______ voluntarily cease to keep or claim; give up

______ an unprincipled person, scoundrel

______ done in a pompous or overly solemn manner

Vocabulary Review Crossword I

Across

4. reserved or uncommunicative in speech; saying little
6. of such excellence as to inspire great admiration
8. to carry lightly through air or over water
9. fluent or persuasive speech or writing
10. entirely lacking or free from

Down

1. sounding sad and mournful
2. honor; the state of being honored
3. to force or urge to do something
5. caution with regard to practical matters
7. morally ignoble or base; vile; dirty or filthy

Vocabulary Review Crossword II

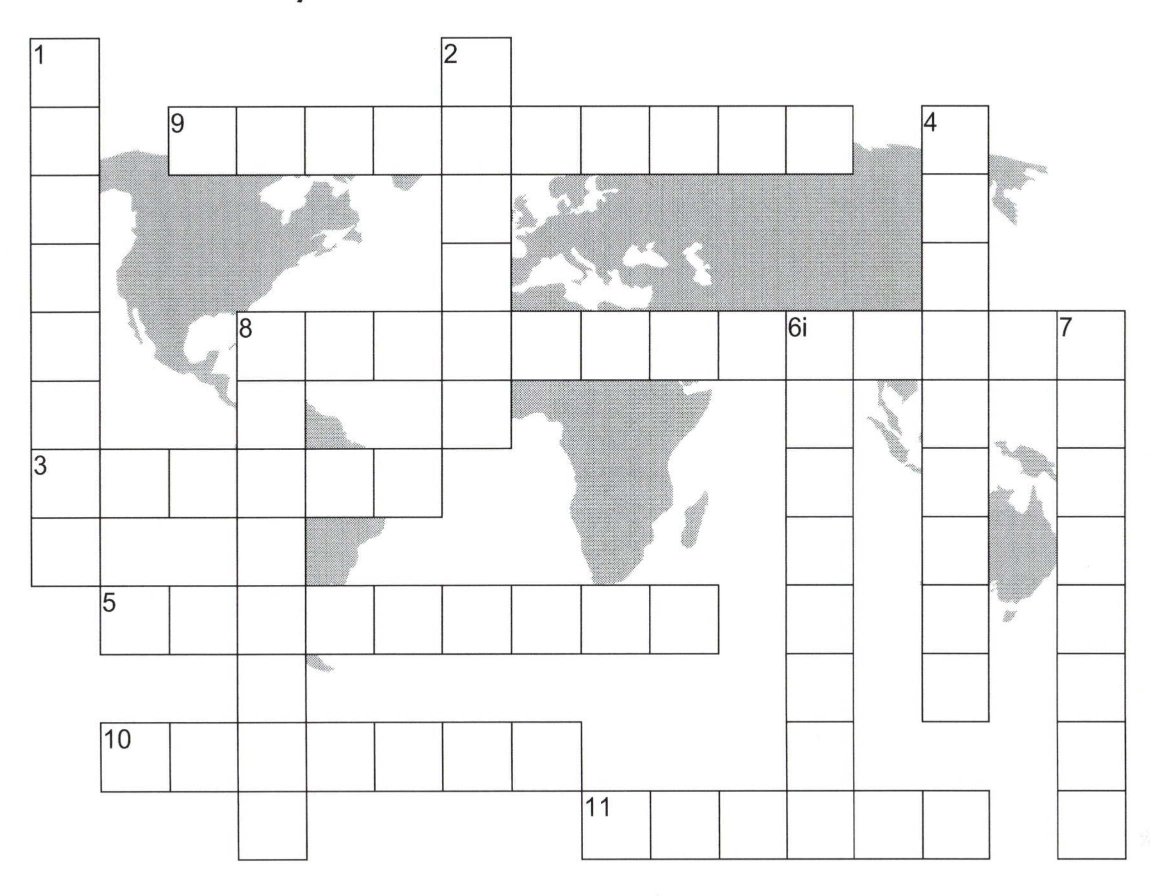

Across

3. show, exhibit
5. a plan or scheme
8. speaking by oneself
9. a means of transportation
10. honor; the state of being honored
11. to carry lightly through air or over water

Down

1. to force or urge to do something
2. entirely lacking or free from
4. sounding sad and mournful
6. offensively bold; impertinent
7. having a harsh, grating sound
8. capable of producing a deep or ringing sound

Vocabulary Review Word Search I

Find the hidden words and explain their meanings. The words can go in any direction, even backwards! (The solution is on page 104.)

C	Z	X	X	E	C	N	A	L	U	T	E	P	E
R	E	S	O	N	A	N	T	D	G	M	N	D	C
D	E	X	T	E	R	I	T	Y	N	O	S	E	N
G	N	I	Z	I	U	Q	O	L	I	L	O	S	E
I	N	T	E	R	J	E	C	T	R	L	D	T	L
Y	N	R	U	T	I	C	A	T	R	I	D	I	U
D	I	D	R	O	S	G	U	G	E	F	X	T	C
E	V	I	T	N	I	A	L	P	N	Y	M	U	U
T	P	O	R	T	E	N	T	O	U	S	A	T	R
F	S	H	I	E	L	O	Q	U	E	N	C	E	T
A	M	M	E	C	N	A	V	E	I	R	G	E	J
W	G	D	E	G	R	E	V	N	O	C	W	Z	B

converged	truculence	sordid	resonant
eloquence	destitute	unerring	taciturn
mitigation	grievance	dexterity	portentous
soliloquizing	mollify	wafted	plaintive

Vocabulary Review Word Search II

Find the hidden words and explain their meanings. The words can go in any direction, even backwards! (The solution is on page 105.)

E	T	A	B	O	R	P	E	R	N	P	D	V	E
A	G	Y	W	E	N	I	S	C	M	R	M	C	C
U	N	O	B	T	R	U	S	I	V	E	L	Y	N
D	I	P	Q	S	L	N	T	R	G	C	A	D	A
A	R	N	R	U	T	I	C	A	T	O	R	E	Y
C	R	W	U	U	G	C	T	B	K	C	D	L	E
I	E	H	S	A	D	A	Y	R	B	I	E	L	V
O	N	R	T	U	R	E	M	A	K	O	N	E	N
U	U	I	E	T	Z	C	N	B	D	U	T	P	O
S	O	L	S	E	V	I	N	C	E	S	L	M	C
N	D	E	S	T	I	T	U	T	E	L	Y	I	W
S	U	O	I	C	I	R	P	A	C	Y	L	K	Y

ardently	evince	conveyance	barbaric
capricious	precociously	impelled	destitute
audacious	sinewy	prudence	mitigation
taciturn	unerring	stratagem	reprobate

Vocabulary Review Word Search III

Find the hidden words and explain their meanings. The words can go in any direction, even backwards! (The solution is on page 105.)

E	C	N	A	N	G	U	P	E	R	I	U	I	I
Q	S	U	O	I	R	E	P	M	I	Z	E	M	U
S	K	K	B	S	E	B	F	I	F	C	M	P	Q
U	T	P	R	U	R	L	W	G	N	S	E	E	P
O	O	R	E	O	M	K	B	E	Z	R	T	N	A
U	R	O	R	U	T	V	D	I	V	S	N	D	L
C	T	G	U	D	V	A	V	A	G	J	E	I	P
O	H	E	C	R	C	N	D	N	U	N	T	N	I
N	O	N	S	A	H	E	P	M	Z	F	A	G	T
N	D	Y	B	F	E	I	G	N	E	D	L	T	A
I	O	G	O	S	U	O	I	R	A	C	I	V	N
U	X	R	P	E	T	A	R	E	C	A	L	Z	T

latent	obscure	vicarious	orthodox
progeny	imperious	innocuous	palpitant
pervade	arduous	impending	feigned
lacerate	cadence	repugnance	tangible

Vocabulary Review Fill-in-the-Blanks I

Fill in the blanks to complete the sentences. Change the word forms if necessary.

plaintively	conveyance	DEXTERITY	latent	
capricious	sordid	evince	soliloquize	fastidious
lacerate	defraud	mitigation	unobtrusively	

1. Nero was a cruel and ________________ emperor.

2. She's always singing some old song ________________.

3. His face ________________ confidence and composure.

4. Shirley was falsely accused of ________________ her clients.

5. The actor in the lead role ________________ often and well.

6. The ________________ of objectionable matter is important.

7. He spoke with the king quietly and ________________ for a while.

8. Barges are used for the ________________ of coal along the Ohio River.

9. The movie described the ____________ history of the mafia since 1950.

10. You need manual ________________ to be good at video games.

11. Some viruses can remain ________________ in the body for years.

12. Her hand was ________________ by a broken glass bottle.

13. The chef is very ________________ about the food he makes.

Vocabulary Review Fill-in-the-Blanks II

Fill in the blanks to complete the sentences. Change the word forms if necessary.

ostentatiously relinquish ARDENTLY
precocious petulance taciturn destitute
converged mortification audacious

1. Elaine studied ________________ to pass the exam.

2. He is reserved and ________________ by nature.

3. I dressed ________________ when I was young.

4. Their next move was even more ________________ and bold.

5. There was a signpost where the two paths ________________.

6. When he died, his family was left completely ________________.

7. Your ________________ made everyone in the room uncomfortable.

8. Esmerelda is ____________ control of her family's business next week.

9. From an early age she displayed a ________________ talent for music.

10. If I could have explained these things to Miss Caroline, I would have saved myself some inconvenience and Miss Caroline's subsequent ________________, but it was beyond my ability to explain things as well as Atticus, so I said, "You are shamin' him, Miss Caroline."

– Harper Lee, *To Kill a Mockingbird*

Vocabulary Review Fill-in-the-Blanks III

Fill in the blanks to complete the sentences. Change the word forms if necessary.

wafted	sublime	impending	DEVOID	
mollify	portentous	commingled	impelled	eloquence
sinewy	indeterminate	unerring	mitigation	

1. Captain Wallace is ________________ in his judgment of the seas.

2. The aroma of bread in the oven ________________ upstairs.

3. Eric's speech was ________________ of warmth and feeling.

4. The art display in the university museum is positively ______________.

5. ________________ of injury risk is the job of our safety officer.

6. The chairman is admired for his intelligence and ________________.

7. With his large frame and ______________ arms, Billy is a great fighter.

8. This author's _______________, self-congratulating tone is annoying.

9. One tiny clue __________ Detective Alexander to investigate further.

10. Will this be enough to ________________ the outraged citizens?

11. Buck had a vague feeling of ________________ doom.

12. Our vacation plan is still at an ________________ stage.

13. The book contained fact ________________ with fiction.

Vocabulary Review Multiple Choice I

Choose the word from the definition or the definition from the word.

voluntarily cease to keep or claim; give up

○ resonant ○ relinquish ○ guttural ○ conveyance

to cut, especially by tearing

○ conciliate ○ interlace ○ lacerate ○ pervade

always right or accurate

○ impudent ○ sulk ○ venture ○ unerring

character trait of someone who loves to talk about himself

○ egotism ○ pertinacity ○ repugnance ○ courier

an unprincipled person, scoundrel

○ stratagem ○ broodings ○ reprobate ○ barbaric

reserved or uncommunicative in speech; saying little

○ rued ○ capricious ○ assure ○ taciturn

converged

○ done in a pompous or overly solemn manner

○ morally ignoble or base; vile; dirty or filthy

○ come together from different directions so as eventually to meet

Vocabulary Review Multiple Choice II

Choose the word from the definition or the definition from the word.

speaking by oneself

○ resonant ○ fathom ○ soliloquizing ○ trajectory

serving in the place of someone or something else

○ latent ○ imperious ○ salient ○ vicarious

having a harsh, grating sound

○ impudent ○ guttural ○ luminary ○ interposed

mixed together

○ commingled ○ tangible ○ palpitant ○ plethoric

show, exhibit

○ evince ○ vaguely ○ interrogative ○ venerable

done in a pompous or overly solemn manner

○ destitute ○ grievance ○ grove ○ portentous

mortification:

○ a feeling of humiliation or shame

○ fluent or persuasive speech or writing

○ of such excellence as to inspire great admiration

Vocabulary Review Multiple Choice III

Choose the word from the definition or the definition from the word.

entirely lacking or free from

○ keen ○ devoid ○ aquatic ○ impelled

following established traditions and beliefs

○ inexorable ○ arduous ○ innocuous ○ orthodox

caution with regard to practical matters

○ prudence ○ mollify ○ sordid ○ sublime

pretended; faked

○ formidable ○ fastidious ○ feigned ○ impending

statement of reasons against an idea or something

○ remonstrance ○ retrogression ○ expediency ○ ministrations

appease the anger or anxiety of (someone)

○ ignorant ○ mollify ○ torment ○ assure

skill in performing tasks, especially with hands

○ adorn ○ forge ○ malicious ○ dexterity

unclear or unable to be determined

○ progeny ○ superficial ○ malinger ○ indeterminate

Synonyms Matching I

Match the words with their synonyms. (The solution is on page 105.)

teeming	comprehend
dilate	disturbed
reverberate	eager
miscellany	piece
agitated	doubtful
fragment	widen
fathom	ask
keen	associate
acquaintance	overflowing
trajectory	hateful
suspicious	resound
adrift	assortment
inquire	afloat
malicious	path

Synonyms Matching II

Match the words with their synonyms (The solution is on page 105.)

tumultuous	lumpy
cumulative	subdue
rut	thanks
rue	ambiguous
knoll	turbulent
gratitude	hurry
knobby	increasing
vague	conflict
inexplicable	unaccountable
repress	hilltop
dismay	satisfy
haste	furrow
strife	distress
quench	regret

Synonyms Multiple Choice

Synonyms are words that have the same or nearly the same meaning. Choose the synonym for each word. (The solution is on page 106.)

command	determined	coax	astonish
○ answer	○ weak	○ persuade	○ expect
○ expect	○ resolute	○ discourage	○ explain
○ respect	○ flexible	○ remove	○ calm
○ order	○ irresolute	○ repel	○ astound

ancient	anticipate	cherish	combine
○ current	○ expect	○ coax	○ negotiate
○ modern	○ forbid	○ adore	○ blend
○ aged	○ combine	○ remove	○ disjoin
○ fresh	○ admire	○ barricade	○ divide

barrier	adjustable	confounded	dainty
○ flexibility	○ flexible	○ clear	○ elegant
○ help	○ determined	○ baffled	○ crude
○ opening	○ appropriate	○ adorable	○ inferior
○ barricade	○ aged	○ available	○ horrible

Antonyms Matching I

Match the words with their antonyms. (The solution is on page 106.)

capricious	apathetic
gratitude	gentle
barbaric	modest
converge	full
ardent	conceal
impel	hinder
impudent	flexible
ferocious	civilized
evince	evade
devoid	disperse
obstinate	encouraged
conflict	steady
confront	harmony
discouraged	ingratitude

Antonyms Matching II

Match the words with their antonyms. (The solution is on page 106.)

DECEPTION	humble
COMPLAINT	appreciative
BOASTFUL	indifferent
PRACTICAL	aggravate
FERVENT	organized
CONSPICUOUS	solution
DESTITUTE	compliment
APPEASE	honesty
UNGRATEFUL	innocent
DISORDERLY	impractical
COMMON	fiction
PROBLEM	affluent
FACT	rare
GUILTY	indistinct

Antonyms Multiple Choice

Antonyms are words that have opposite or nearly opposite meanings. Choose the antonym for each word. (The solution is on page 106.)

appropriate	generous	drought	agony
O improper	O benevolent	O scarcity	O comfort
O relevant	O lavish	O leak	O barrier
O correct	O stingy	O comfort	O anguish
O useful	O accurate	O flood	O torment

discourage	malfunction	consent	achievement
O eliminate	O function	O approval	O realization
O adore	O defect	O permit	O blend
O inspire	O bug	O blessing	O defeat
O comfort	O glitch	O denial	O feat

absorb	forbid	deliberately	accurate
O leak	O ban	O accidentally	O distinct
O command	O allow	O purposely	O scientific
O swallow	O discourage	O willfully	O skillful
O consent	O order	O knowingly	O inexact

Solutions to Selected Activities

Page 21
Level 5 Vocabulary Review I

immense
indispensable
venerable
contradict
derision
ambition
sulk
indignant

Page 22
Level 6 Vocabulary Review II

dilating
trajectory
profanation
interposed
acquainted
baritone

making a loud, confused noise;
uproarious

Page 23
Level 5 Vocabulary Review II

vaguely
languid
piteous
detriment
perversity
irrepressible
economize
cultivate

Page 30
Analogies Matching I

toe : foot :: finger : hand
eyes : see :: teeth : chew
stem : flower :: trunk : tree
fish : gills :: human : lungs
cold : ice :: heat : fire
pig : pork :: cow : beef

Page 31
Analogies Matching II

pencil : write :: knife : cut
hungry : eat :: tired : rest
glasses : see :: cane : walk
boring : interesting :: rare : common
key : unlock :: scale : weigh
cub : bear :: kitten : cat

Page 32
Analogies Matching III

match : fire :: key : car
candy : sweet :: fire : hot
teacher : educate :: doctor : heal
whale : ocean :: lion : jungle
television : watch : novel : read
brother : sibling :: father : parent

Page 33

Treasure Island Trivia I

Jim Hawkins
The Admiral Benbow Inn
John
illness
an island
Spy-glass Hill
Porto Bello
Robert Louise Stevenson

Page 34

Treasure Island Trivia II

Savannah
Barbecue
The Hispanolia
Captain Flint
a cook
Billy Bones
Jolly Roger
Pieces of eight

Page 37

Level 6 Vocabulary Review II

luminary
flecked
keen
subdue
fathom
miscellany
fragmentary
gnarled

Page 44

Penrod and Sam I Word Search

The first letters are marked. Remember that the words can go in any direction!

S	J	R	A	E	M	R	D	Q	V
X	G	H	E	R	C	T	C	G	E
G	M	N	I	H	Y	V	H	L	V
S	U	T	I	U	K	I	J	B	I
O	F	T	Q	D	P	C	A	A	T
N	Q	L	T	D	O	R	H	T	N
O	V	A	H	U	B	O	N	L	I
R	O	O	U	A	R	E	R	T	A
O	N	G	R	G	D	A	D	B	L
U	N	I	Z	U	A	E	L	L	P
S	C	M	P	Z	I	J	U	Z	N
C	W	M	J	X	U	L	W	V	Z
G	I	R	E	S	O	N	A	N	T
N	C	V	G	I	N	J	L	I	R
L	T	B	Y	O	Z	Z	S	B	K

Page 50

Penrod and Sam II Word Search

E	F	E	R	O	C	I	T	Y	P
C	K	C	D	K	O	M	E	L	L
N	X	N	C	P	N	U	N	S	A
E	H	A	E	Z	V	N	S	U	I
L	Y	V	M	Q	E	O	T	O	N
U	T	E	O	C	R	B	R	I	T
C	I	I	L	K	G	T	A	C	I
U	R	R	L	I	E	R	T	O	V
R	E	G	I	L	D	U	A	C	E
T	T	F	F	R	D	S	G	E	L
F	X	C	Y	E	V	I	E	R	Y
L	E	E	V	O	K	V	M	P	U
E	D	O	X	N	J	E	C	J	X
B	I	M	Y	F	K	L	V	C	N
D	O	S	W	F	K	Y	K	D	L
D	X	Y	U	G	Z	K	E	L	M

Page 73

Treasure Island Characters

C. Long John Silver
G. Jim Hawkins
B. Captain Smollet
I. Mr. Arrow
H. Dr. Livesey
L. Israel Hands
J. O'Brien
E. Squire Trelawney
D. Ben Gunn
K. Blind Pew
F. Billy Bones
A. Black Dog

Page 74

Treasure Island Word Search I

The first letters are marked. Remember that the words can go in any direction!

Page 75

Treasure Island Word Search II

The first letters are marked. Remember that the words can go in any direction!

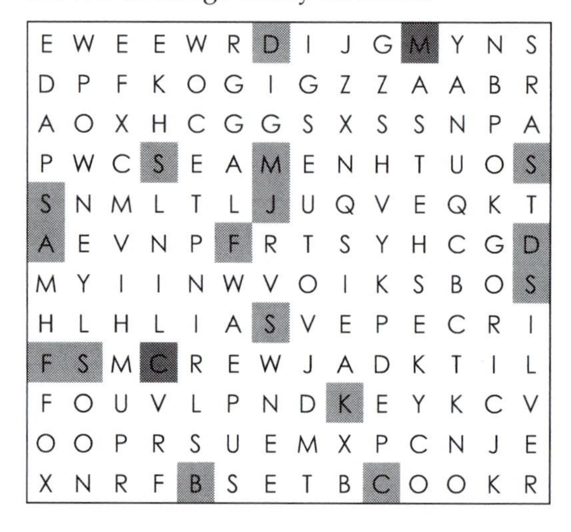

Page 86

Vocabulary Review Word Search I

The first letters are marked. Remember that the words can go in any direction!

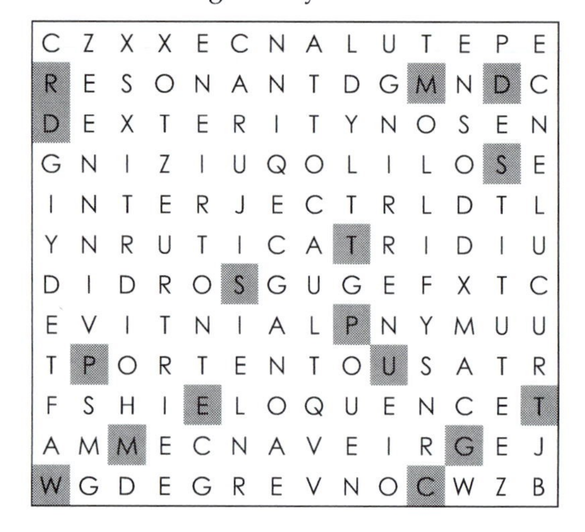

Page 87

Vocabulary Review Word Search II

The first letters are marked. Remember that the words can go in any direction!

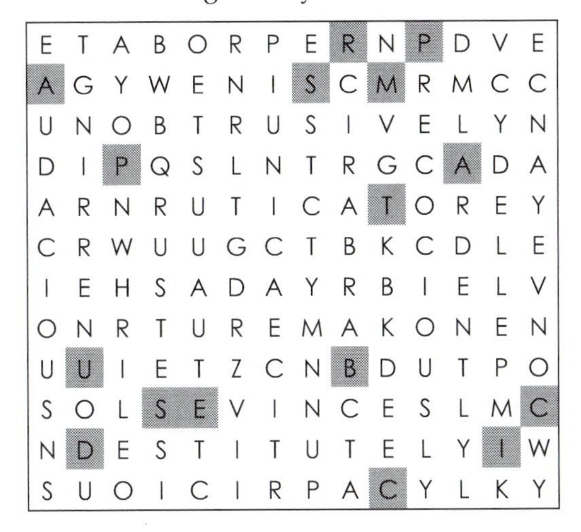

E	T	A	B	O	R	P	E	R	N	P	D	V	E
A	G	Y	W	E	N	I	S	C	M	R	M	C	C
U	N	O	B	T	R	U	S	I	V	E	L	Y	N
D	I	P	Q	S	L	N	T	R	G	C	A	D	A
A	R	N	R	U	T	I	C	A	T	O	R	E	Y
C	R	W	U	U	G	C	T	B	K	C	D	L	E
I	E	H	S	A	D	A	Y	R	B	I	E	L	V
O	N	R	T	U	R	E	M	A	K	O	N	E	N
U	U	I	E	T	Z	C	N	B	D	U	T	P	O
S	O	L	S	E	V	I	N	C	E	S	L	M	C
N	D	E	S	T	I	T	U	T	E	L	Y	I	W
S	U	O	I	C	I	R	P	A	C	Y	L	K	Y

Page 88

Vocabulary Review Word Search III

The first letters are marked. Remember that the words can go in any direction!

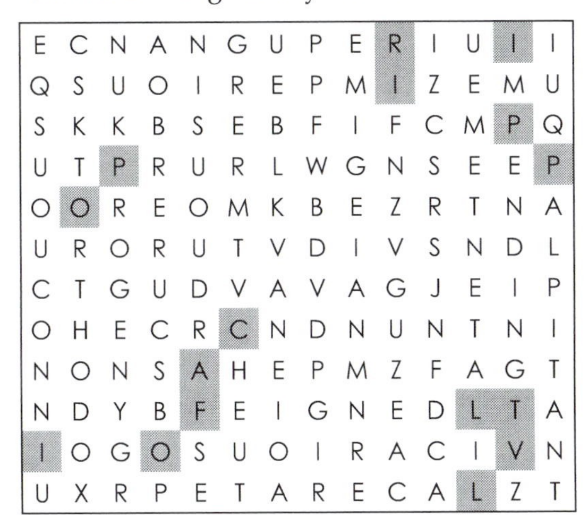

E	C	N	A	N	G	U	P	E	R	I	U	I	I
Q	S	U	O	I	R	E	P	M	I	Z	E	M	U
S	K	K	B	S	E	B	F	I	F	C	M	P	Q
U	T	P	R	U	R	L	W	G	N	S	E	E	P
O	O	R	E	O	M	K	B	E	Z	R	T	N	A
U	R	O	R	U	T	V	D	I	V	S	N	D	L
C	T	G	U	D	V	A	V	A	G	J	E	I	P
O	H	E	C	R	C	N	D	N	U	N	T	N	I
N	O	N	S	A	H	E	P	M	Z	F	A	G	T
N	D	Y	B	F	E	I	G	N	E	D	L	T	A
I	O	G	O	S	U	O	I	R	A	C	I	V	N
U	X	R	P	E	T	A	R	E	C	A	L	Z	T

Page 95

Synonyms Matching I

teeming – overflowing
dilate – widen
reverberate – resound
miscellany – assortment
agitated – disturbed
fragment – piece
fathom – comprehend
keen – eager
acquaintance – associate
trajectory – path
suspicious – doubtful
adrift – afloat
inquire – ask
malicious – hateful

Page 96

Synonyms Matching II

tumultuous – turbulent
cumulative – increasing
rut – furrow
rue – regret
knoll – hilltop
gratitude – thanks
knobby – lumpy
vague – ambiguous
inexplicable – unaccountable
repress – subdue
dismay – distress
haste – hurry
strife – conflict
quench – satisfy

Page 97

Synonyms Multiple Choice

command - order
determined - resolute
coax - persuade
astonish - astound

ancient - aged
anticipate - expect
cherish - adore
combine - blend

barrier - barricade
adjustable - flexible
confounded - baffled
dainty - elegant

Page 98

Antonyms Matching I

capricious – steady
gratitude – ingratitude
barbaric – civilized
converge – disperse
ardent – apathetic
impel – hinder
impudent – modest
ferocious – gentle
evince – conceal
devoid – full
obstinate – flexible
conflict- harmony
confront – evade
discouraged – encouraged

Page 99

Antonyms Matching II

deception – honesty
complaint – compliment
boastful – humble
practical – impractical
fervent – indifferent
conspicuous – indistinct
destitute – affluent
appease – aggravate
ungrateful – appreciative
disorderly – organized
common – rare
problem – solution
fact – fiction
guilty – innocent

Page 100

Antonyms Multiple Choice

appropriate - improper
generous - stingy
drought - flood
agony - comfort

discourage - inspire
malfunction - function
consent - denial
achievement - defeat

absorb - leak
forbid - allow
deliberately - accidentally
accurate - inexact

Made in the USA
Coppell, TX
31 August 2020

35863964R00059

Chapter 15 Exercises – Part B

Directions: Perform the following integrals using integration by parts.

❸ $\int \frac{\ln x}{x^2} dx =$

❹ $\int_{x=0}^{\pi/6} \sin x \tan x \, dx =$

❖ Check your answers at the back of the book.

Chapter 15 Exercises – Part C

Directions: Perform the following integrals using integration by parts.

❺ $\int x^2 \cos x \, dx =$

❻ $\int e^x \sin x \, dx =$

❖ Check your answers at the back of the book.

16 MULTIPLE INTEGRALS

To perform a double or triple integral, follow these steps:

1. Note that you may reverse the order of the differential elements. Don't worry about whether you see $dxdy$ or $dydx$. The order of the differential elements doesn't tell you which integral to do first.
2. Look at the limits of integration: If you see an integration variable (like x, y, or z) in the limits of integration, you must perform the integral with the variable limit first. You will see variable limits in three of the examples that follow.
3. When you integrate over one variable, treat the other independent variables as if they are constants. For example, when performing an integral over the variable x, treat y as a constant. Similarly, when performing an integral over the variable y, treat x as a constant. We will see this in the examples.
4. After you finish one integral, evaluate its antiderivative over the limits before you begin the next integral.

Example: Perform the following integral.

$$\int_{x=0}^{2} \int_{y=0}^{x^2} xy \, dx \, dy$$

We must integrate over y first because the integral over y has a variable limit (x^2). When we integrate over y, we treat the independent variable x as a constant. Pull x out of the y integral (but be careful not to pull x out of the x integral).

$$\int_{x=0}^{2} \int_{y=0}^{x^2} xy \, dx \, dy = \int_{x=0}^{2} x \left(\int_{y=0}^{x^2} y \, dy \right) dx = \int_{x=0}^{2} x \left[\frac{y^2}{2} \right]_{y=0}^{x^2} dx = \frac{1}{2} \int_{x=0}^{2} x [y^2]_{y=0}^{x^2} \, dx$$

$$= \frac{1}{2} \int_{x=0}^{2} x[(x^2)^2 - 0^2] \, dx = \frac{1}{2} \int_{x=0}^{2} x x^4 \, dx = \frac{1}{2} \int_{x=0}^{2} x^5 \, dx = \frac{1}{2} \left[\frac{x^6}{6} \right]_{x=0}^{2} = \frac{1}{12} [x^6]_{x=0}^{2}$$

$$= \frac{1}{12}(2^6 - 0^6) = \frac{1}{12}(64) = \frac{64}{12} = \frac{64 \div 4}{12 \div 4} = \frac{16}{3} \approx 5.333$$

Example: Perform the following integral.

$$\int_{x=0}^{1/y}\int_{y=1}^{3} xy^2\,dx\,dy$$

We must integrate over x first because the integral over x has a variable limit $(1/y)$. When we integrate over x, we treat the independent variable y as a constant. Pull y out of the x integral (but be careful not to pull y out of the y integral).

$$\int_{x=0}^{1/y}\int_{y=1}^{3} xy^2\,dx\,dy = \int_{y=1}^{3} y^2\left(\int_{x=0}^{1/y} x\,dx\right)dy = \int_{y=1}^{3} y^2\left[\frac{x^2}{2}\right]_{x=0}^{1/y} dy = \frac{1}{2}\int_{y=1}^{3} y^2[x^2]_{x=0}^{1/y}\,dy$$

$$= \frac{1}{2}\int_{y=1}^{3} y^2\left[\left(\frac{1}{y}\right)^2 - 0^2\right]dy = \frac{1}{2}\int_{y=1}^{3}\frac{y^2}{y^2}dy = \frac{1}{2}\int_{y=1}^{3} dy = \frac{1}{2}[y]_{y=1}^{3} = \frac{1}{2}(3-1) = \frac{2}{2} = 1$$

Example: Perform the following integral.

$$\int_{x=0}^{1/3}\int_{y=0}^{3} x^2y^2\,dx\,dy$$

Since all of the limits are constants, we may perform these integrals in any order.

$$\int_{x=0}^{1/3}\int_{y=0}^{3} x^2y^2\,dx\,dy = \int_{x=0}^{1/3} x^2\left(\int_{y=0}^{3} y^2\,dy\right)dx = \int_{x=0}^{1/3} x^2\left[\frac{y^3}{3}\right]_{y=0}^{3} dx$$

$$= \frac{1}{3}\int_{x=0}^{1/3} x^2[y^3]_{y=0}^{3}\,dx = \frac{1}{3}\int_{x=0}^{1/3} x^2(3^3 - 0^3)\,dx = \frac{1}{3}\int_{x=0}^{1/3} x^2(27)\,dx = 9\int_{x=0}^{1/3} x^2\,dx$$

$$= 9\left[\frac{x^3}{3}\right]_{x=0}^{1/3} = 3[x^3]_{x=0}^{1/3} = 3\left[\left(\frac{1}{3}\right)^3 - 0^3\right] = 3\left(\frac{1}{27}\right) = \frac{3}{27} = \frac{3\div 3}{27\div 3} = \frac{1}{9} \approx 0.111$$

Example: Perform the following integral.

$$\int_{x=0}^{1}\int_{y=0}^{x^2}\int_{z=0}^{y} xyz\,dx\,dy\,dz$$

We must integrate over y and z before x because the integrals over y and z have variable limits (x^2 and y). Furthermore, we must integrate over z before y because the limit in the z integration includes y. We will first integrate over z, then integrate over y, and lastly integrate over x.

When we integrate over z, we treat the independent variables x and y as constants. Pull x and y out of the z integral (but be careful not to pull y out of the y integral).

$$\int_{x=0}^{1}\int_{y=0}^{x^2}\int_{z=0}^{y} xyz\,dx\,dy\,dz = \int_{x=0}^{1} x \int_{y=0}^{x^2} y\left(\int_{z=0}^{y} z\,dz\right)dy\,dx = \int_{x=0}^{1} x \int_{y=0}^{x^2} y\left[\frac{z^2}{2}\right]_{z=0}^{y} dy\,dx$$

$$= \frac{1}{2}\int_{x=0}^{1} x \int_{y=0}^{x^2} y[z^2]_{z=0}^{y}\,dy\,dx = \frac{1}{2}\int_{x=0}^{1} x \int_{y=0}^{x^2} y(y^2-0)\,dy\,dx = \frac{1}{2}\int_{x=0}^{1} x \int_{y=0}^{x^2} yy^2\,dy\,dx$$

$$= \frac{1}{2}\int_{x=0}^{1} x \int_{y=0}^{x^2} y^3\,dy\,dx = \frac{1}{2}\int_{x=0}^{1} x\left(\int_{y=0}^{x^2} y^3\,dy\right)dx = \frac{1}{2}\int_{x=0}^{1} x\left[\frac{y^4}{4}\right]_{y=0}^{x^2} dx$$

$$= \frac{1}{8}\int_{x=0}^{1} x[y^4]_{y=0}^{x^2}\,dx = \frac{1}{8}\int_{x=0}^{1} x[(x^2)^4-0^4]\,dx = \frac{1}{8}\int_{x=0}^{1} x(x^8-0)\,dx = \frac{1}{8}\int_{x=0}^{1} xx^8\,dx$$

$$= \frac{1}{8}\int_{x=0}^{1} x^9\,dx = \frac{1}{8}\left[\frac{x^{10}}{10}\right]_{x=0}^{1} = \frac{1}{80}[x^{10}]_{x=0}^{1} = \frac{1}{80}(1^{10}-0^{10})$$

$$= \frac{1}{80}(1-0) = \frac{1}{80} = 0.0125$$

Note that $(x^2)^4 = x^8$ according to the rule $(x^m)^n = x^{mn}$.

Chapter 16 Exercises – Part A

Directions: Perform the following integrals.

❶ $$\int_{x=0}^{3} \int_{y=0}^{\sqrt{x}} xy\, dx\, dy =$$

❷ $$\int_{x=0}^{y^2} \int_{y=1}^{2} \frac{y^2}{\sqrt{x}} dx\, dy =$$

❖ Check your answers at the back of the book.

Chapter 16 Exercises – Part B

Directions: Perform the following integrals.

❸ $$\int_{x=1}^{4}\int_{y=4}^{9}\frac{dxdy}{\sqrt{xy}} =$$

❹ $$\int_{x=0}^{5}\int_{y=x}^{2x} x^2y\,dx\,dy =$$

❖ Check your answers at the back of the book.

Chapter 16 Exercises – Part C

Directions: Perform the following integrals.

❺ $$\int_{x=0}^{y}\int_{y=0}^{2}\int_{z=0}^{x} xy^2z^3\,dx\,dy\,dz =$$

❻ $$\int_{x=0}^{\sqrt{y}}\int_{y=0}^{\sqrt{z}}\int_{z=0}^{4} x^3y\,dx\,dy\,dz =$$

❖ Check your answers at the back of the book.

SOLUTIONS

Chapter 1, Part A

❶ $\frac{d}{dx}(8x^4) = (4)(8)x^{4-1} = \boxed{32x^3}$

❷ $\frac{d}{dx}(5x^{-2}) = (-2)(5)x^{-2-1} = -10x^{-3} = \boxed{-\frac{10}{x^3}}$

Note: Both $-10x^{-3}$ and $-\frac{10}{x^3}$ are correct answers.

❸ $\frac{d}{dt}\left(\frac{1}{t}\right) = \frac{d}{dt}(t^{-1}) = \frac{d}{dt}(1t^{-1}) = (-1)(1)t^{-1-1} = -t^{-2} = \boxed{-\frac{1}{t^2}}$

Note: Both $-t^{-2}$ and $-\frac{1}{t^2}$ are correct answers.

❹ $\frac{d}{dx}(8x^{7/4}) = \left(\frac{7}{4}\right)(8)x^{7/4-1} = \frac{56}{4}x^{3/4} = \boxed{14x^{3/4}}$

Note: $\frac{7}{4} - 1 = \frac{7}{4} - \frac{4}{4} = \frac{7-4}{4} = \frac{3}{4}$ (subtract fractions with a common denominator).

❺ $\frac{d}{dx}\left(\frac{x^{3/5}}{6}\right) = \left(\frac{3}{5}\right)\left(\frac{1}{6}\right)x^{3/5-1} = \frac{3}{30}x^{-2/5} = \frac{x^{-2/5}}{10} = \boxed{\frac{1}{10x^{2/5}}}$

Notes:

- $\frac{3}{5} - 1 = \frac{3}{5} - \frac{5}{5} = \frac{3-5}{5} = \frac{-2}{5}$ (subtract fractions with a common denominator).
- Both $\frac{x^{-2/5}}{10}$ and $\frac{1}{10x^{2/5}}$ are correct answers.

❻ $\frac{d}{du}(u) = \frac{d}{du}(1u^1) = (1)(1)u^{1-1} = 1u^0 = \boxed{1}$

Notes: $1u^1 = u$ and $u^0 = 1$.

❼ $\frac{d}{dt}\left(\sqrt{2t}\right) = \frac{d}{dt}(2t)^{1/2} = \frac{d}{dt}\left(2^{1/2}t^{1/2}\right) = \left(\frac{1}{2}\right)\left(2^{1/2}\right)t^{1/2-1} = \frac{2^{1/2}}{2}t^{-1/2}$

$= 2^{1/2-1}t^{-1/2} = 2^{-1/2}t^{-1/2} = \frac{1}{2^{1/2}t^{1/2}} = \frac{1}{(2t)^{1/2}} = \frac{1}{\sqrt{2t}} = \frac{1}{\sqrt{2t}}\frac{\sqrt{2t}}{\sqrt{2t}} = \boxed{\frac{\sqrt{2t}}{2t}}$

Notes:

- $(2t)^{1/2} = 2^{1/2}t^{1/2}$ because $(cx)^n = c^n x^n$.
- $\left(\frac{1}{2}\right)\left(2^{1/2}\right) = \frac{2^{1/2}}{2} = 2^{1/2}2^{-1} = 2^{1/2-1} = 2^{-1/2}$ because $x^m x^n = x^{m+n}$.
- $\frac{1}{2} - 1 = \frac{1}{2} - \frac{2}{2} = \frac{1-2}{2} = -\frac{1}{2}$ (subtract fractions with a common denominator).
- $2^{-1/2}t^{-1/2}$, $\frac{1}{2^{1/2}t^{1/2}}$, $\frac{1}{\sqrt{2t}}$, and $\frac{\sqrt{2t}}{2t}$ are all correct answers. However, only the answer $\frac{\sqrt{2t}}{2t}$ has a rational denominator. We multiplied $\frac{1}{\sqrt{2t}}$ by $\frac{\sqrt{2t}}{\sqrt{2t}}$ in order to rationalize the denominator. Note that $\sqrt{2t}\sqrt{2t} = 2t$ because $\sqrt{x}\sqrt{x} = x$.

❽ $\frac{d}{dx}\left(\frac{1}{\sqrt{x}}\right) = \frac{d}{dx}\left(\frac{1}{x^{1/2}}\right) = \frac{d}{dx}\left(x^{-1/2}\right) = \frac{d}{dx}\left(1x^{-1/2}\right) = \left(-\frac{1}{2}\right)(1)x^{-1/2-1}$

$= \left(-\frac{1}{2}\right)(1)x^{-3/2} = -\frac{x^{-3/2}}{2} = -\frac{1}{2x^{3/2}} = -\frac{1}{2x\sqrt{x}} = -\frac{1}{2x\sqrt{x}}\frac{\sqrt{x}}{\sqrt{x}} = \boxed{-\frac{\sqrt{x}}{2x^2}}$

Notes:

- $\sqrt{x} = x^{1/2}$
- $\frac{1}{x^{1/2}} = x^{-1/2}$ because $x^{-n} = \frac{1}{x^n}$.
- $x^{3/2} = x^1 x^{1/2} = x\sqrt{x}$ because $x^{m+n} = x^m x^n$.
- $\sqrt{x}\sqrt{x} = x$.
- $-\frac{x^{-3/2}}{2}$, $-\frac{1}{2x^{3/2}}$, $-\frac{1}{2x\sqrt{x}}$, and $-\frac{\sqrt{x}}{2x^2}$ are all correct answers. However, only the answer $-\frac{\sqrt{x}}{2x^2}$ has a rational denominator. We multiplied $-\frac{1}{2x\sqrt{x}}$ by $\frac{\sqrt{x}}{\sqrt{x}}$ in order to rationalize the denominator.

Chapter 1, Part B

❾ $\frac{d}{dx}(5x^3+4x^2-3x+2) = \frac{d}{dx}(5x^3) + \frac{d}{dx}(4x^2) - \frac{d}{dx}(3x) + \frac{d}{dx}(2)$

$= (3)(5)x^{3-1} + (2)(4)x^{2-1} - (1)(3)x^{1-1} + 0 = \boxed{15x^2+8x-3}$

Notes:

- $\frac{d}{dx}(3x) = \frac{d}{dx}(3x^1) = (1)(3)x^{1-1} = 3x^0 = 3$ because $x^1 = x$ and $x^0 = 1$.
- $\frac{d}{dx}(2) = 0$ because the derivative of a constant is zero.

❿ $\frac{d}{du}(1-u) = \frac{d}{du}(1) - \frac{d}{du}(u) = 0 - 1 = \boxed{-1}$

Notes:

- $\frac{d}{dx}(1) = 0$ because the derivative of a constant is zero.
- $\frac{d}{du}(u) = \frac{d}{du}(1u^1) = (1)(1)u^{1-1} = 1u^0 = 1$ because $1u^1 = u$ and $u^0 = 1$.

⓫ $\frac{d}{dx}\left(4x^{3/2}+12x^{1/2}\right) = \left(\frac{3}{2}\right)(4)x^{3/2-1} + \left(\frac{1}{2}\right)(12)x^{1/2-1} = \frac{12}{2}x^{1/2} + \frac{12}{2}x^{-1/2}$

$= 6x^{1/2} + 6x^{-1/2} = 6x^{1/2} + \frac{6}{x^{1/2}} = 6\sqrt{x} + \frac{6}{\sqrt{x}} = 6\sqrt{x} + \frac{6}{\sqrt{x}}\frac{\sqrt{x}}{\sqrt{x}} = \boxed{6\sqrt{x} + \frac{6\sqrt{x}}{x}}$

Notes:

- $\frac{3}{2} - 1 = \frac{3}{2} - \frac{2}{2} = \frac{3-2}{2} = \frac{1}{2}$ (subtract fractions with a common denominator).
- $\frac{1}{2} - 1 = \frac{1}{2} - \frac{2}{2} = \frac{1-2}{2} = -\frac{1}{2}$ (subtract fractions with a common denominator).
- $x^{-1/2} = \frac{1}{x^{1/2}}$ because $x^{-n} = \frac{1}{x^n}$.
- $x^{1/2} = \sqrt{x}$.
- $6x^{1/2} + 6x^{-1/2}$, $6x^{1/2} + \frac{6}{x^{1/2}}$, $6\sqrt{x} + \frac{6}{\sqrt{x}}$, and $6\sqrt{x} + \frac{6\sqrt{x}}{x}$ are all correct answers.

  However, only the answer $6\sqrt{x} + \frac{6\sqrt{x}}{x}$ has a rational denominator. We multiplied $\frac{6}{\sqrt{x}}$ by $\frac{\sqrt{x}}{\sqrt{x}}$ in order to rationalize the denominator.

⓬ $\frac{d}{dt}\left(\sqrt{t}-\frac{1}{t}\right)=\frac{d}{dt}\left(t^{1/2}-t^{-1}\right)=\frac{d}{dt}\left(1t^{1/2}-1t^{-1}\right)$

$$=\left(\frac{1}{2}\right)(1)t^{1/2-1}-(-1)(1)t^{-1-1}=\frac{t^{-1/2}}{2}-(-t^{-2})=\frac{t^{-1/2}}{2}+t^{-2}=\frac{1}{2t^{1/2}}+t^{-2}$$

$$=\frac{1}{2\sqrt{t}}+\frac{1}{t^2}=\frac{1}{2\sqrt{t}}\frac{\sqrt{t}}{\sqrt{t}}+\frac{1}{t^2}=\boxed{\frac{\sqrt{t}}{2t}+\frac{1}{t^2}}$$

Notes:

- $\sqrt{t}=t^{1/2}$ and $\frac{1}{t}=t^{-1}$.
- $-(-t^{-2})=t^{-2}$ (two negatives make a positive).
- $\frac{1}{2}-1=\frac{1}{2}-\frac{2}{2}=\frac{1-2}{2}=-\frac{1}{2}$ (subtract fractions with a common denominator).
- $-1-1=-2$.
- $\frac{t^{-1/2}}{2}+t^{-2}$, $\frac{1}{2t^{1/2}}+t^{-2}$, $\frac{1}{2\sqrt{t}}+\frac{1}{t^2}$, and $\frac{\sqrt{t}}{2t}+\frac{1}{t^2}$ are all correct answers. However, only the answer $\frac{\sqrt{t}}{2t}+\frac{1}{t^2}$ has a rational denominator. We multiplied $\frac{1}{2\sqrt{t}}$ by $\frac{\sqrt{t}}{\sqrt{t}}$ in order to rationalize the denominator.

Chapter 2, Part A

❶ $\frac{d}{dx}(x^3-3x^2+4x-5)^8=?$

Apply the chain rule: $u=x^3-3x^2+4x-5$, $f=u^8$, $\frac{df}{dx}=?$

$$\frac{df}{dx}=\frac{df}{du}\frac{du}{dx}=\left[\frac{d}{du}(u^8)\right]\left[\frac{d}{dx}(x^3-3x^2+4x-5)\right]$$

$$=(8u^7)(3x^2-6x+4)=\boxed{8(x^3-3x^2+4x-5)^7(3x^2-6x+4)}$$

❷ $\frac{d}{dt}\frac{1}{\sqrt{5t^2-3t+6}}=?$

Apply the chain rule: $u=5t^2-3t+6$, $f=\frac{1}{\sqrt{u}}=\frac{1}{u^{1/2}}=u^{-1/2}$, $\frac{df}{dt}=?$

$$\frac{df}{dt}=\frac{df}{du}\frac{du}{dt}=\left[\frac{d}{du}\left(u^{-1/2}\right)\right]\left[\frac{d}{dt}(5t^2-3t+6)\right]=\left(-\frac{1}{2}u^{-3/2}\right)(10t-3)$$

$$=-\frac{10t-3}{2u^{3/2}}=\boxed{-\frac{10t-3}{2(5t^2-3t+6)^{3/2}}}$$

Notes: $-\frac{1}{2}-1=-\frac{1}{2}-\frac{2}{2}=\frac{-1-2}{2}=-\frac{3}{2}$ and $u^{-3/2}=\frac{1}{u^{3/2}}$.

❸ $\frac{d}{dx}(4x^2 - 6)\sqrt{x} = ?$

Apply the product rule: $f(x) = 4x^2 - 6$, $g(x) = \sqrt{x} = x^{1/2}$, $\frac{d}{dx}(fg) = ?$

$$\frac{d}{dx}(fg) = g\frac{df}{dx} + f\frac{dg}{dx} = x^{1/2}\left[\frac{d}{dx}(4x^2 - 6)\right] + (4x^2 - 6)\left(\frac{d}{dx}x^{1/2}\right)$$

$$= x^{1/2}(8x) + (4x^2 - 6)\left(\frac{1}{2}x^{-1/2}\right) = 8x^{1/2}x + 2x^2x^{-1/2} - 3x^{-1/2}$$

$$= 8x^{3/2} + 2x^{3/2} - 3x^{-1/2} = 10x^{3/2} - 3x^{-1/2} = 10x^{3/2} - \frac{3}{x^{1/2}} = 10x^{3/2} - \frac{3}{\sqrt{x}}$$

$$= 10x^{3/2} - \frac{3\sqrt{x}}{x} = \boxed{10x\sqrt{x} - \frac{3\sqrt{x}}{x}}$$

Notes:

- $\frac{1}{2} - 1 = \frac{1}{2} - \frac{2}{2} = \frac{1-2}{2} = -\frac{1}{2}$ (subtract fractions with a common denominator).
- $x^{1/2}x = x^{1/2}x^1 = x^{3/2}$ because $x^1 = x$ and $x^m x^n = x^{m+n}$.
- $x^2x^{-1/2} = x^{3/2}$ because $x^p x^{-q} = x^{p-q}$.
- Going from $(4x^2 - 6)\left(\frac{1}{2}x^{-1/2}\right)$ to $2x^2x^{-1/2} - 3x^{-1/2}$, distribute according to $(r + s)t = rt + st$ with $r = 4x^2$, $s = -6$, and $t = \frac{1}{2}x^{-1/2}$.
- $10x^{3/2} - 3x^{-1/2}$, $10x^{3/2} - \frac{3}{x^{1/2}}$, $10x^{3/2} - \frac{3}{\sqrt{x}}$, $10x^{3/2} - \frac{3\sqrt{x}}{x}$, and $10x\sqrt{x} - \frac{3\sqrt{x}}{x}$ are all correct answers. We multiplied $\frac{3}{\sqrt{x}}$ by $\frac{\sqrt{x}}{\sqrt{x}}$ in order to rationalize the denominator. In the last step, we used $x^{3/2} = x^1x^{1/2} = xx^{1/2} = x\sqrt{x}$.

❹ $\frac{d}{dx}\frac{3 - 2x^2}{4 - 3x^2} = ?$

Apply the quotient rule: $f(x) = 3 - 2x^2$, $g(x) = 4 - 3x^2$, $\frac{d}{dx}\left(\frac{f}{g}\right) = ?$

$$\frac{d}{dx}\left(\frac{f}{g}\right) = \frac{g\frac{df}{dx} - f\frac{dg}{dx}}{g^2} = \frac{(4 - 3x^2)\left[\frac{d}{dx}(3 - 2x^2)\right] - (3 - 2x^2)\left[\frac{d}{dx}(4 - 3x^2)\right]}{(4 - 3x^2)^2}$$

$$= \frac{(4 - 3x^2)(-4x) - (3 - 2x^2)(-6x)}{16 - 12x^2 - 12x^2 + 9x^4} = \frac{-16x + 12x^3 + 18x - 12x^3}{16 - 24x^2 + 9x^4}$$

$$= \boxed{\frac{2x}{16 - 24x^2 + 9x^4}}$$

Notes: $-(3)(-6x) = 18x$ and $-(-2x^2)(-6x) = -12x^3$.

Chapter 2, Part B

❺ $\frac{d}{dx}\left(\frac{1}{x^3 - 4x}\right) = ?$

Apply the chain rule: $u = x^3 - 4x$, $f = \frac{1}{u} = u^{-1}$, $\frac{df}{dx} = ?$

$$\frac{df}{dx} = \frac{df}{du}\frac{du}{dx} = \left[\frac{d}{du}(u^{-1})\right]\left[\frac{d}{dx}(x^3 - 4x)\right] = (-u^{-2})(3x^2 - 4) = -\frac{(3x^2 - 4)}{u^2}$$

$$= \frac{-(3x^2 - 4)}{(x^3 - 4x)^2} = \frac{-3x^2 - (-4)}{(x^3 - 4x)^2} = \boxed{\frac{-3x^2 + 4}{(x^3 - 4x)^2}}$$

Note that $\frac{d}{du}(u^{-1}) = -u^{-2}$ because $-1 - 1 = -2$.

❻ $\frac{d}{dx}(4 + x)^9(2 - x)^5 = ?$

Apply the product rule: $f(x) = (4 + x)^9$, $g(x) = (2 - x)^5$, $\frac{d}{dx}(fg) = ?$

$$\frac{d}{dx}(fg) = g\frac{df}{dx} + f\frac{dg}{dx} = (2 - x)^5\left[\frac{d}{dx}(4 + x)^9\right] + (4 + x)^9\left[\frac{d}{dx}(2 - x)^5\right]$$

Now apply the chain rule: $u = 4 + x$, $f = u^9$, $v = 2 - x$, $g = v^5$

$$\frac{d}{dx}(fg) = (2 - x)^5\left[\frac{df}{du}\frac{du}{dx}\right] + (4 + x)^9\left[\frac{dg}{dv}\frac{dv}{dx}\right]$$

$$= (2 - x)^5\left[\frac{d}{du}(u^9)\frac{d}{dx}(4 + x)\right] + (4 + x)^9\left[\frac{d}{dv}(v^5)\frac{d}{dx}(2 - x)\right]$$

$$= (2 - x)^5[(9u^8)(1)] + (4 + x)^9[(5v^4)(-1)] = 9u^8(2 - x)^5 - 5v^4(4 + x)^9$$

$$= \boxed{9(4 + x)^8(2 - x)^5 - 5(4 + x)^9(2 - x)^4}$$

Although $9(4 + x)^8(2 - x)^5 - 5(4 + x)^9(2 - x)^4$ is a correct answer, if you want to be fancy, you could factor out $(4 + x)^8(2 - x)^4$ as follows:

$$9(4 + x)^8(2 - x)^5 - 5(4 + x)^9(2 - x)^4 = (4 + x)^8(2 - x)^4[9(2 - x) - 5(4 + x)]$$

$$= (4 + x)^8(2 - x)^4(18 - 9x - 20 - 5x) = (4 + x)^8(2 - x)^4(-2 - 14x)$$

$$= (4 + x)^8(2 - x)^4(-1)(2 + 14x) = \boxed{-(4 + x)^8(2 - x)^4(2 + 14x)}$$

Note:

- $-5(4 + x) = -5(4) - 5(x) = -20 - 5x$ (the minus sign gets distributed).
- $(-2 - 14x) = (-1)(2 + 14x) = -(2 + 14x)$, where again a minus sign gets distributed.

7 $\frac{d}{dt}\left(\sqrt{t^2+9}\right) = ?$

Apply the chain rule: $u = t^2 + 9$, $f = \sqrt{u} = u^{1/2}$, $\frac{df}{dt} = ?$

$$\frac{df}{dt} = \frac{df}{du}\frac{du}{dt} = \left[\frac{d}{du}\left(u^{1/2}\right)\right]\left[\frac{d}{dt}(t^2+9)\right] = \left(\frac{1}{2}u^{-1/2}\right)(2t) = \frac{2t}{2u^{1/2}} = \frac{t}{\sqrt{u}} = \frac{t}{\sqrt{t^2+9}}$$

$$= \frac{t}{\sqrt{t^2+9}}\frac{\sqrt{t^2+9}}{\sqrt{t^2+9}} = \boxed{\frac{t\sqrt{t^2+9}}{t^2+9}}$$

Notes:

- $\frac{1}{2} - 1 = \frac{1}{2} - \frac{2}{2} = \frac{1-2}{2} = -\frac{1}{2}$ (subtract fractions with a common denominator).
- $\frac{t}{\sqrt{t^2+9}}$ and $\frac{t\sqrt{t^2+9}}{t^2+9}$ are both correct, but $\frac{t\sqrt{t^2+9}}{t^2+9}$ has a rational denominator.

8 $\frac{d}{dx}\frac{x^4}{\sqrt{x^2+4}} = ?$

Apply the quotient rule: $f(x) = x^4$, $g(x) = \sqrt{x^2+4}$, $\frac{d}{dx}\left(\frac{f}{g}\right) = ?$

$$\frac{d}{dx}\left(\frac{f}{g}\right) = \frac{g\frac{df}{dx} - f\frac{dg}{dx}}{g^2} = \frac{\sqrt{x^2+4}\left[\frac{d}{dx}x^4\right] - x^4\left[\frac{d}{dx}\sqrt{x^2+4}\right]}{\left(\sqrt{x^2+4}\right)^2} = \frac{\sqrt{x^2+4}(4x^3) - x^4\left(\frac{d}{dx}\sqrt{x^2+4}\right)}{x^2+4}$$

Now apply the chain rule: $u = x^2 + 4$, $g = \sqrt{u} = u^{1/2}$

$$\frac{4x^3\sqrt{x^2+4} - x^4\left(\frac{dg}{du}\frac{du}{dx}\right)}{x^2+4} = \frac{4x^3\sqrt{x^2+4} - x^4\left[\frac{d}{du}\left(u^{1/2}\right)\frac{d}{dx}(x^2+4)\right]}{x^2+4}$$

$$= \frac{4x^3\sqrt{x^2+4} - x^4\left[\left(\frac{1}{2}u^{-1/2}\right)(2x)\right]}{x^2+4} = \frac{4x^3\sqrt{x^2+4} - x^5u^{-1/2}}{x^2+4}$$

$$= \frac{4x^3\sqrt{x^2+4} - x^5(x^2+4)^{-1/2}}{x^2+4} = 4x^3(x^2+4)^{-1/2} - x^5(x^2+4)^{-3/2}$$

$$= \frac{4x^3}{(x^2+4)^{1/2}} - \frac{x^5}{(x^2+4)^{3/2}} = \frac{4x^3(x^2+4) - x^5}{(x^2+4)^{3/2}} = \boxed{\frac{3x^5+16x^3}{(x^2+4)^{3/2}}}$$

Notes:

- In the second to last line, we distributed: $\frac{A+B}{C} = \frac{A}{C} + \frac{B}{C} = AC^{-1} + BC^{-1}$.
- $\sqrt{x^2+4}(x^2+4)^{-1} = (x^2+4)^{1/2}(x^2+4)^{-1} = (x^2+4)^{-1/2}$ according to $x^m x^n = x^{m+n}$. Similarly, $(x^2+4)^{-1/2}(x^2+4)^{-1} = (x^2+4)^{-3/2}$.
- $\frac{4x^3}{(x^2+4)^{1/2}} = \frac{4x^3}{(x^2+4)^{1/2}}\frac{(x^2+4)}{(x^2+4)} = \frac{4x^3(x^2+4)}{(x^2+4)^{3/2}} = \frac{4x^5+16x^3}{(x^2+4)^{3/2}}$.

Chapter 2, Part C

9 $\frac{d}{dx}\left(2x^{5/2} - 8x^{3/2}\right)^6 = ?$

Apply the chain rule: $u = 2x^{5/2} - 8x^{3/2}$, $f = u^6$, $\frac{df}{dx} = ?$

$$\frac{df}{dx} = \frac{df}{du}\frac{du}{dx} = \left[\frac{d}{du}(u^6)\right]\left[\frac{d}{dx}(2x^{5/2} - 8x^{3/2})\right] = (6u^5)\left[\left(\frac{5}{2}\right)(2)x^{3/2} - \left(\frac{3}{2}\right)(8)x^{1/2}\right]$$

$$= (6u^5)\left(5x^{3/2} - 12x^{1/2}\right) = \boxed{6\left(2x^{5/2} - 8x^{3/2}\right)^5\left(5x^{3/2} - 12x^{1/2}\right)}$$

10 $\frac{d}{dx}\frac{x^2 + 3x - 4}{2x + 5} = ?$

Apply the quotient rule: $f(x) = x^2 + 3x - 4$, $g(x) = 2x + 5$, $\frac{d}{dx}\left(\frac{f}{g}\right) = ?$

$$\frac{d}{dx}\left(\frac{f}{g}\right) = \frac{g\frac{df}{dx} - f\frac{dg}{dx}}{g^2} = \frac{(2x+5)\left[\frac{d}{dx}(x^2 + 3x - 4)\right] - (x^2 + 3x - 4)\left[\frac{d}{dx}(2x+5)\right]}{(2x+5)^2}$$

$$= \frac{(2x+5)(2x+3) - (x^2 + 3x - 4)(2)}{4x^2 + 10x + 10x + 25} = \frac{4x^2 + 6x + 10x + 15 - 2x^2 - 6x + 8}{4x^2 + 20x + 25}$$

$$= \boxed{\frac{2x^2 + 10x + 23}{4x^2 + 20x + 25}}$$

Note: Distribute the minus sign in $-(x^2 + 3x - 4)(2)$ to get $(-x^2 - 3x + 4)(2)$.

11 $\frac{d}{dx}\sqrt{2 + \sqrt{x}} = ?$

Apply the chain rule: $u = 2 + \sqrt{x} = 2 + x^{1/2}$, $f = \sqrt{u} = u^{1/2}$, $\frac{df}{dx} = ?$

$$\frac{df}{dx} = \frac{df}{du}\frac{du}{dx} = \left[\frac{d}{du}(u^{1/2})\right]\left[\frac{d}{dx}(2 + x^{1/2})\right] = \left(\frac{1}{2}u^{-1/2}\right)\left(\frac{1}{2}x^{-1/2}\right) = \frac{1}{4}u^{-1/2}x^{-1/2}$$

$$= \frac{1}{4u^{1/2}x^{1/2}} = \frac{1}{4\sqrt{u}\sqrt{x}} = \frac{1}{4\sqrt{x}\sqrt{u}} = \frac{1}{4\sqrt{x}\sqrt{2 + \sqrt{x}}} = \frac{1}{4\sqrt{x(2 + \sqrt{x})}} = \frac{1}{4\sqrt{2x + x\sqrt{x}}}$$

$$= \frac{1}{4\sqrt{2x + x\sqrt{x}}}\frac{\sqrt{2x + x\sqrt{x}}}{\sqrt{2x + x\sqrt{x}}} = \frac{\sqrt{2x + x\sqrt{x}}}{4(2x + x\sqrt{x})} = \frac{\sqrt{2x + x\sqrt{x}}}{4(2x + x\sqrt{x})}\frac{(2x - x\sqrt{x})}{(2x - x\sqrt{x})}$$

$$= \frac{(2x - x\sqrt{x})\sqrt{2x + x\sqrt{x}}}{4(4x^2 - x^2\sqrt{x}\sqrt{x})} = \boxed{\frac{(2x - x\sqrt{x})\sqrt{2x + x\sqrt{x}}}{4(4x^2 - x^3)}}$$

(any of the last 8 steps is a correct answer)

⓬ $\frac{d}{dt}(4t^2 - 9)(t^4 + 8t^2 - 3)^9 = ?$

Apply the product rule: $f(t) = 4t^2 - 9$, $g(t) = (t^4 + 8t^2 - 3)^9$, $\frac{d}{dt}(fg) = ?$

$$\frac{d}{dt}(fg) = g\frac{df}{dt} + f\frac{dg}{dt}$$

$$= (t^4 + 8t^2 - 3)^9\left[\frac{d}{dt}(4t^2 - 9)\right] + (4t^2 - 9)\left[\frac{d}{dt}(t^4 + 8t^2 - 3)^9\right]$$

$$= (t^4 + 8t^2 - 3)^9(8t) + (4t^2 - 9)\left[\frac{d}{dt}(t^4 + 8t^2 - 3)^9\right]$$

Now apply the chain rule: $u = t^4 + 8t^2 - 3$, $g = u^9$

$$\frac{d}{dt}(fg) = (8t)(t^4 + 8t^2 - 3)^9 + (4t^2 - 9)\left[\frac{dg}{du}\frac{du}{dt}\right]$$

$$= (8t)(t^4 + 8t^2 - 3)^9 + (4t^2 - 9)\left[\frac{d}{du}(u^9)\frac{d}{dt}(t^4 + 8t^2 - 3)\right]$$

$$= (8t)(t^4 + 8t^2 - 3)^9 + (4t^2 - 9)[(9u^8)(4t^3 + 16t)]$$

$$= (8t)(t^4 + 8t^2 - 3)^9 + 9u^8(4t^3 + 16t)(4t^2 - 9)$$

$$= \boxed{(8t)(t^4 + 8t^2 - 3)^9 + 9(t^4 + 8t^2 - 3)^8(4t^3 + 16t)(4t^2 - 9)}$$

Although $(8t)(t^4 + 8t^2 - 3)^9 + 9(t^4 + 8t^2 - 3)^8(4t^3 + 16t)(4t^2 - 9)$ is a correct answer, if you want to be fancy, you could factor out $(t^4 + 8t^2 - 3)^8$ as follows:

$$(8t)(t^4 + 8t^2 - 3)^9 + 9(t^4 + 8t^2 - 3)^8(4t^3 + 16t)(4t^2 - 9)$$

$$= (t^4 + 8t^2 - 3)^8[(8t)(t^4 + 8t^2 - 3) + 9(4t^3 + 16t)(4t^2 - 9)]$$

$$= (t^4 + 8t^2 - 3)^8[8t^5 + 64t^3 - 24t + 9(16t^5 - 36t^3 + 64t^3 - 144t)]$$

$$= (t^4 + 8t^2 - 3)^8[8t^5 + 64t^3 - 24t + 9(16t^5 + 28t^3 - 144t)]$$

$$= (t^4 + 8t^2 - 3)^8(8t^5 + 64t^3 - 24t + 144t^5 + 252t^3 - 1296t)$$

$$= \boxed{(t^4 + 8t^2 - 3)^8(152t^5 + 316t^3 - 1320t)}$$

Chapter 3, Part A

❶ $\frac{d}{d\theta}7\tan 5\theta$

Apply the chain rule: $u = 5\theta$, $f = 7\tan u$, $\frac{df}{d\theta} = ?$

$$\frac{df}{d\theta} = \frac{df}{du}\frac{du}{d\theta} = \left[\frac{d}{du}(7\tan u)\right]\left[\frac{d}{d\theta}(5\theta)\right] = (7\sec^2 u)(5) = 35\sec^2 u = \boxed{35\sec^2(5\theta)}$$

Note: As usual, the constant coefficient simply comes out: $\frac{d}{du}(7\tan u) = 7\frac{d}{du}(\tan u)$.

❷ $\frac{d}{d\theta}3\sin^4\theta$

Apply the chain rule: $u = \sin\theta$, $f = 3u^4$, $\frac{df}{d\theta} = ?$

$$\frac{df}{d\theta} = \frac{df}{du}\frac{du}{d\theta} = \left[\frac{d}{du}(3u^4)\right]\left[\frac{d}{d\theta}(\sin\theta)\right] = (12u^3)(\cos\theta) = \boxed{12\sin^3\theta\cos\theta}$$

Note: It is instructive to compare this solution to the previous solution.

❸ $\frac{d}{d\theta}\csc\theta\sec\theta$

Apply the product rule: $f(\theta) = \csc\theta$, $g(\theta) = \sec\theta$, $\frac{d}{d\theta}(fg) = ?$

$$\frac{d}{d\theta}(fg) = g\frac{df}{d\theta} + f\frac{dg}{d\theta} = \sec\theta\left(\frac{d}{d\theta}\csc\theta\right) + \csc\theta\left(\frac{d}{d\theta}\sec\theta\right)$$

$$= \sec\theta\,(-\csc\theta\cot\theta) + \csc\theta\,(\sec\theta\tan\theta) = -\sec\theta\csc\theta\cot\theta + \csc\theta\sec\theta\tan\theta$$

$$= \sec\theta\csc\theta\,(-\cot\theta + \tan\theta) = \frac{1}{\cos\theta\sin\theta}\left(-\frac{\cos\theta}{\sin\theta} + \frac{\sin\theta}{\cos\theta}\right)$$

$$= \frac{-1}{\cos\theta\sin\theta}\frac{\cos\theta}{\sin\theta} + \frac{1}{\cos\theta\sin\theta}\frac{\sin\theta}{\cos\theta} = -\frac{1}{\sin^2\theta} + \frac{1}{\cos^2\theta} = \boxed{-\csc^2\theta + \sec^2\theta}$$

Notes:

- $\csc\theta = \frac{1}{\sin\theta}$, $\sec\theta = \frac{1}{\cos\theta}$, $\tan\theta = \frac{\sin\theta}{\cos\theta}$, and $\cot\theta = \frac{\cos\theta}{\sin\theta}$.
- $\sec\theta\csc\theta\,(-\cot\theta + \tan\theta)$ and $-\frac{1}{\sin^2\theta} + \frac{1}{\cos^2\theta}$, are also correct answers.

❹ $\frac{d}{d\theta}\frac{\sin\theta + \cos\theta}{\sin\theta}$

Apply the quotient rule: $f(\theta) = \sin\theta + \cos\theta$, $g(\theta) = \sin\theta$, $\frac{d}{d\theta}\left(\frac{f}{g}\right) = ?$

$$\frac{d}{d\theta}\left(\frac{f}{g}\right) = \frac{g\frac{df}{d\theta} - f\frac{dg}{d\theta}}{g^2} = \frac{\sin\theta\left[\frac{d}{d\theta}(\sin\theta + \cos\theta)\right] - (\sin\theta + \cos\theta)\left(\frac{d}{d\theta}\sin\theta\right)}{\sin^2\theta}$$

$$= \frac{\sin\theta\,(\cos\theta - \sin\theta) - (\sin\theta + \cos\theta)(\cos\theta)}{\sin^2\theta} = \frac{\sin\theta\cos\theta - \sin^2\theta - \sin\theta\cos\theta - \cos^2\theta}{\sin^2\theta}$$

$$= \frac{-\sin^2\theta - \cos^2\theta}{\sin^2\theta} = -\frac{1}{\sin^2\theta} = \boxed{-\csc^2\theta}$$

Notes:

- $\frac{d}{d\theta}(\sin\theta + \cos\theta) = \frac{d}{d\theta}\sin\theta + \frac{d}{d\theta}\cos\theta = \cos\theta - \sin\theta$.
- $\sin^2\theta + \cos^2\theta = 1$ and $\csc\theta = \frac{1}{\sin\theta}$.

Chapter 3, Part B

❺ $\frac{d}{d\theta}\cos(\theta^2 - 2\theta + 4)$

Apply the chain rule: $u = \theta^2 - 2\theta + 4$, $f = \cos u$, $\frac{df}{d\theta} = ?$

$$\frac{df}{d\theta} = \frac{df}{du}\frac{du}{d\theta} = \left[\frac{d}{du}(\cos u)\right]\left[\frac{d}{d\theta}(\theta^2 - 2\theta + 4)\right] = (-\sin u)(2\theta - 2)$$
$$= -(2\theta - 2)\sin(\theta^2 - 2\theta + 4) = \boxed{-2(\theta - 1)\sin(\theta^2 - 2\theta + 4)}$$

Note that $2\theta - 2 = 2(\theta - 1)$. We factored out the 2.

❻ $\frac{d}{d\theta}2\cot\sqrt{\theta}$

Apply the chain rule: $u = \sqrt{\theta} = \theta^{1/2}$, $f = 2\cot u$, $\frac{df}{d\theta} = ?$

$$\frac{df}{d\theta} = \frac{df}{du}\frac{du}{d\theta} = \left[\frac{d}{du}(2\cot u)\right]\left[\frac{d}{d\theta}(\theta^{1/2})\right] = (-2\csc^2 u)\left(\frac{1}{2}\theta^{-1/2}\right)$$
$$= -2\csc^2 u\left(\frac{1}{2\theta^{1/2}}\right) = -\frac{2\csc^2 u}{2\theta^{1/2}} = -\frac{\csc^2 u}{\theta^{1/2}} = -\frac{\csc^2\sqrt{\theta}}{\sqrt{\theta}} = -\left(\frac{\csc^2\sqrt{\theta}}{\sqrt{\theta}}\right)\frac{\sqrt{\theta}}{\sqrt{\theta}}$$
$$= \boxed{-\frac{\sqrt{\theta}\csc^2\sqrt{\theta}}{\theta}} = -\frac{\sqrt{\theta}}{\theta\sin^2\sqrt{\theta}}$$

Notes:

- $\theta^{-1/2} = \frac{1}{\theta^{1/2}}$ according to $x^{-m} = \frac{1}{x^m}$.
- $\sqrt{\theta}\sqrt{\theta} = \theta$.
- $-\frac{\csc^2\sqrt{\theta}}{\theta^{1/2}}$, $-\frac{\csc^2\sqrt{\theta}}{\sqrt{\theta}}$, and $-\frac{\sqrt{\theta}\csc^2\sqrt{\theta}}{\theta}$ are all correct answers. However, only the answers $-\frac{\sqrt{\theta}\csc^2\sqrt{\theta}}{\theta}$ and $-\frac{\sqrt{\theta}}{\theta\sin^2\sqrt{\theta}}$ have a rational denominator. We multiplied $-\frac{\csc^2\sqrt{\theta}}{\sqrt{\theta}}$ by $\frac{\sqrt{\theta}}{\sqrt{\theta}}$ in order to rationalize the denominator.
- $\csc\theta = \frac{1}{\sin\theta}$.
- Note that $\csc^2\sqrt{\theta}$ doesn't simplify to $\csc\theta$.

❼ $\frac{d}{d\theta}(\theta \sin\theta)$

Apply the product rule: $f(\theta) = \theta$, $g(\theta) = \sin\theta$, $\frac{d}{d\theta}(fg) = ?$

$$\frac{d}{d\theta}(fg) = g\frac{df}{d\theta} + f\frac{dg}{d\theta} = \sin\theta\left(\frac{d}{d\theta}\theta\right) + \theta\left(\frac{d}{d\theta}\sin\theta\right) = (\sin\theta)(1) + \theta(\cos\theta)$$

$$= \boxed{\sin\theta + \theta\cos\theta}$$

Note that $\frac{d}{d\theta}\theta = 1$ just as $\frac{d}{dx}x = 1$. In more steps, $\frac{d}{dx}(1x^1) = (1)(1)x^{1-1} = x^0 = 1$.

❽ $\frac{d}{d\theta}\sqrt{1 + \sin\theta}$

Apply the chain rule: $u = 1 + \sin\theta$, $f = \sqrt{u} = u^{1/2}$, $\frac{df}{d\theta} = ?$

$$\frac{df}{d\theta} = \frac{df}{du}\frac{du}{d\theta} = \left[\frac{d}{du}\left(u^{1/2}\right)\right]\left[\frac{d}{d\theta}(1 + \sin\theta)\right] = \left(\frac{1}{2}u^{-1/2}\right)(0 + \cos\theta)$$

$$= \left(\frac{1}{2u^{1/2}}\right)\cos\theta = \frac{\cos\theta}{2u^{1/2}} = \frac{\cos\theta}{2\sqrt{u}} = \frac{\cos\theta}{2\sqrt{1 + \sin\theta}} = \frac{\cos\theta}{2\sqrt{1 + \sin\theta}}\frac{\sqrt{1 + \sin\theta}}{\sqrt{1 + \sin\theta}} = \boxed{\frac{\cos\theta\sqrt{1 + \sin\theta}}{2(1 + \sin\theta)}}$$

Chapter 3, Part C

❾ $\frac{d}{dx}3\cot^{-1}(4x)$

Apply the chain rule: $u = 4x$, $f = 3\cot^{-1}u$, $\frac{df}{dx} = ?$

$$\frac{df}{dx} = \frac{df}{du}\frac{du}{dx} = \left[\frac{d}{du}(3\cot^{-1}u)\right]\left[\frac{d}{dx}(4x)\right] = \left(\frac{-3}{1 + u^2}\right)(4) = \frac{-12}{1 + u^2}$$

$$= \frac{-12}{1 + (4x)^2} = \boxed{\frac{-12}{1 + 16x^2}}$$

Note that $(4x)^2 = 4^2x^2 = 16x^2$ according to $(ax)^n = a^n x^n$.

❿ $\frac{d}{dx}x\csc^{-1}x$

Apply the product rule: $f(x) = x$, $g(x) = \csc^{-1}x$, $\frac{d}{dx}(fg) = ?$

$$\frac{d}{dx}(fg) = g\frac{df}{dx} + f\frac{dg}{dx} = \csc^{-1}x\left(\frac{d}{dx}x\right) + x\left(\frac{d}{dx}\csc^{-1}x\right)$$

$$= (\csc^{-1}x)(1) + x\left(\frac{-1}{|x|\sqrt{x^2 - 1}}\right) = \boxed{\csc^{-1}x - \frac{x}{|x|\sqrt{x^2 - 1}}} \text{ provided that } |x| > 1$$

⓫ $\frac{d}{dx}(\sin^{-1} x + \cos^{-1} x) = \frac{d}{dx}\sin^{-1} x + \frac{d}{dx}\cos^{-1} x = \frac{1}{\sqrt{1-x^2}} + \frac{-1}{\sqrt{1-x^2}} = \boxed{0}$

⓬ $\frac{d}{dx}\frac{\sec^{-1} x}{x}$

Apply the quotient rule: $f(x) = \sec^{-1} x$, $g(x) = x$, $\frac{d}{dx}\left(\frac{f}{g}\right) = ?$

$$\frac{d}{dx}\left(\frac{f}{g}\right) = \frac{g\frac{df}{dx} - f\frac{dg}{dx}}{g^2} = \frac{x\left[\frac{d}{dx}(\sec^{-1} x)\right] - (\sec^{-1} x)\left(\frac{d}{dx}x\right)}{x^2}$$

$$= \frac{x\left(\frac{1}{|x|\sqrt{x^2-1}}\right) - (\sec^{-1} x)(1)}{x^2} = \frac{\frac{x}{|x|\sqrt{x^2-1}} - \sec^{-1} x}{x^2}$$

$$= \left(\frac{x}{|x|\sqrt{x^2-1}} - \sec^{-1} x\right)\left(\frac{1}{x^2}\right) = \boxed{\frac{1}{x|x|\sqrt{x^2-1}} - \frac{\sec^{-1} x}{x^2}} \text{ where } |x| > 1$$

Note that $\frac{x}{x^2} = \frac{1}{x}$. Also, note the absolute values on $|x|$ (the absolute values in $\frac{d}{dx}\sec^{-1} x = \frac{1}{|x|\sqrt{x^2-1}}$ reflect that a graph of secant inverse has a positive slope for all possible values of both $x > 1$ and $x < 1$).

Chapter 4, Part A

❶ $\frac{d}{dx}\left(4e^{x^2} - 6e^{4x} + 9\right) = \frac{d}{dx}\left(4e^{x^2}\right) - \frac{d}{dx}(6e^{4x}) + \frac{d}{dx}9$

$$= 8xe^{x^2} - 24e^{4x} + 0 = \boxed{8xe^{x^2} - 24e^{4x}}$$

Notes:

- $\frac{d}{dx}\left(4e^{x^2}\right) = \frac{d}{dx}(4e^u)$ where $u = x^2$. Apply the chain rule with $f = 4e^u$ and $u = x^2$: $\frac{df}{dx} = \frac{df}{du}\frac{du}{dx} = \left[\frac{d}{du}(4e^u)\right]\left(\frac{d}{dx}x^2\right) = (4e^u)(2x) = 8xe^u = 8xe^{x^2}$. Therefore, $\frac{d}{dx}\left(4e^{x^2}\right) = 8xe^{x^2}$.
- $\frac{d}{dx}(6e^{4x}) = 6\frac{d}{dx}(e^{4x}) = 6(4e^{4x}) = 24e^{4x}$ according to $\frac{d}{dx}e^{ax} = ae^{ax}$.
- A derivative of a constant is zero: $\frac{d}{dx}(9) = 0$.

❷ $\frac{d}{dx} 4 \cosh^3 x$

Apply the chain rule: $u = \cosh x$, $f = 4u^3$, $\frac{df}{dx} = ?$

$$\frac{df}{dx} = \frac{df}{du}\frac{du}{dx} = \left[\frac{d}{du}(4u^3)\right]\left(\frac{d}{dx}\cosh x\right) = (12u^2)(\sinh x) = \boxed{12 \cosh^2 x \sinh x}$$

Note that the derivative of hyperbolic cosine, $\frac{d}{dx}(\cosh x) = \sinh x$, is positive, whereas the derivative of ordinary cosine, $\frac{d}{dx}\cos x = -\sin x$, is negative.

❸ $\frac{d}{dt} \sinh t \cosh t$

Apply the product rule: $f(t) = \sinh t$, $g(t) = \cosh t$, $\frac{d}{dt}(fg) = ?$

$$\frac{d}{dt}(fg) = g\frac{df}{dt} + f\frac{dg}{dt} = \cosh t\left(\frac{d}{dt}\sinh t\right) + \sinh t\left(\frac{d}{dt}\cosh t\right)$$
$$= \cosh t\,(\cosh t) + \sinh t\,(\sinh t) = \boxed{\cosh^2 t + \sinh^2 t}$$

Notes:

- The answer doesn't reduce to 1. Compare the hyperbolic identity, $\cosh^2 x - \sinh^2 x = 1$, to the ordinary trig identity, $\cos^2 x + \sin^2 x = 1$: Note the minus sign in the hyperbolic identity.
- Note that the derivative of hyperbolic cosine, $\frac{d}{dx}(\cosh x) = \sinh x$, is positive, whereas the derivative of ordinary cosine, $\frac{d}{dx}\cos x = -\sin x$, is negative.

❹ $\frac{d}{dx} \sinh[\cosh(x)]$

This is not a product of hyperbolic functions. Rather, hyperbolic cosine is inside the argument of hyperbolic sine. This is a chain rule problem, not a product rule.

Apply the chain rule: $u = \cosh x$, $f = \sinh u$, $\frac{df}{dx} = ?$

$$\frac{df}{dx} = \frac{df}{du}\frac{du}{dx} = \left(\frac{d}{du}\sinh u\right)\left(\frac{d}{dx}\cosh x\right)$$
$$= (\cosh u)(\sinh x) = \sinh x \cosh u = \boxed{\sinh x \cosh[\cosh(x)]}$$

Note that $\cosh[\cosh(x)]$ has $\cosh(x)$ inside of its argument. This isn't multiplication. Also, note that the derivative of hyperbolic cosine, $\frac{d}{dx}(\cosh x) = \sinh x$, is positive, whereas the derivative of ordinary cosine, $\frac{d}{dx}\cos x = -\sin x$, is negative.

Chapter 4, Part B

❺ $\frac{d}{dx}\tanh\sqrt{x}$

Apply the chain rule: $u = \sqrt{x} = x^{1/2}$, $f = \tanh u$, $\frac{df}{dx} = ?$

$$\frac{df}{dx} = \frac{df}{du}\frac{du}{dx} = \left(\frac{d}{du}\tanh u\right)\left(\frac{d}{dx}x^{1/2}\right) = (\operatorname{sech}^2 u)\left(\frac{1}{2}x^{-1/2}\right) = \frac{1}{2}x^{-1/2}\operatorname{sech}^2 u$$

$$= \frac{\operatorname{sech}^2 u}{2x^{1/2}} = \frac{\operatorname{sech}^2 u}{2\sqrt{x}} = \frac{\operatorname{sech}^2 u}{2\sqrt{x}}\frac{\sqrt{x}}{\sqrt{x}} = \frac{\sqrt{x}\operatorname{sech}^2 u}{2x} = \frac{\sqrt{x}\operatorname{sech}^2\sqrt{x}}{2x} = \boxed{\frac{\sqrt{x}}{2x\cosh^2\sqrt{x}}}$$

Notes:

- $\frac{1}{2} - 1 = \frac{1}{2} - \frac{2}{2} = \frac{1-2}{2} = -\frac{1}{2}$ (subtract fractions with a common denominator).
- $\frac{1}{x^{1/2}} = x^{-1/2}$ because $x^{-n} = \frac{1}{x^n}$.
- $\sqrt{x}\sqrt{x} = x$.
- $\frac{1}{2}x^{-1/2}\operatorname{sech}^2\sqrt{x}$, $\frac{\operatorname{sech}^2\sqrt{x}}{2x^{1/2}}$, $\frac{\operatorname{sech}^2\sqrt{x}}{2\sqrt{x}}$, $\frac{\sqrt{x}\operatorname{sech}^2\sqrt{x}}{2x}$, and $\frac{\sqrt{x}}{2x\cosh^2\sqrt{x}}$ are all correct answers. However, only the last two answers have a rational denominator. We multiplied $\frac{\operatorname{sech}^2\sqrt{x}}{2\sqrt{x}}$ by $\frac{\sqrt{x}}{\sqrt{x}}$ in order to rationalize the denominator.
- $\operatorname{sech} x = \frac{1}{\cosh x}$.
- Note that $\cosh^2\sqrt{x}$ doesn't simplify to $\cosh x$.

❻ $\frac{d}{dt}(t^2e^t)$

Apply the product rule: $f(t) = t^2$, $g(t) = e^t$, $\frac{d}{dt}(fg) = ?$

$$\frac{d}{dt}(fg) = g\frac{df}{dt} + f\frac{dg}{dt} = e^t\left(\frac{d}{dt}t^2\right) + t^2\left(\frac{d}{dt}e^t\right) = e^t(2t) + t^2(e^t) = 2te^t + t^2e^t$$

$$= e^t(2t + t^2) = \boxed{e^t(t^2 + 2t)}$$

Notes:

- $\frac{d}{dt}e^t = e^t$ according to $\frac{d}{dx}e^{ax} = ae^{ax}$ with $a = 1$ and x replaced by t.
- $2te^t + t^2e^t = e^t(2t + t^2)$. Factor out e^t.
- $2t + t^2 = t^2 + 2t$ since addition is commutative (order doesn't matter).

❼ $\frac{d}{dx}\sqrt{1+e^{-x}}$

Apply the chain rule: $u = 1 + e^{-x}$, $f = \sqrt{u} = u^{1/2}$, $\frac{df}{dx} = ?$

$$\frac{df}{dx} = \frac{df}{du}\frac{du}{dx} = \left(\frac{d}{du}u^{1/2}\right)\left[\frac{d}{dx}(1+e^{-x})\right] = \left(\frac{1}{2}u^{-1/2}\right)(0 - e^{-x}) = -\frac{u^{-1/2}e^{-x}}{2}$$

$$= -\frac{e^{-x}}{2u^{1/2}} = -\frac{e^{-x}}{2\sqrt{u}} = -\frac{e^{-x}}{2\sqrt{1+e^{-x}}} = -\frac{e^{-x}}{2\sqrt{1+e^{-x}}}\frac{\sqrt{1+e^{-x}}}{\sqrt{1+e^{-x}}} = \boxed{-\frac{e^{-x}\sqrt{1+e^{-x}}}{2(1+e^{-x})}}$$

Notes:

- $\frac{1}{2} - 1 = \frac{1}{2} - \frac{2}{2} = \frac{1-2}{2} = -\frac{1}{2}$ (subtract fractions with a common denominator).
- $\frac{1}{u^{1/2}} = u^{-1/2}$ because $x^{-n} = \frac{1}{x^n}$.
- A derivative of a constant is zero: $\frac{d}{dx}(1) = 0$.
- $\frac{d}{dx}e^{-x} = -e^{-x}$ according to $\frac{d}{dx}e^{ax} = ae^{ax}$ with $a = -1$.
- $\sqrt{1+e^{-x}}\sqrt{1+e^{-x}} = 1 + e^{-x}$ because $\sqrt{u}\sqrt{u} = u$.
- $-\frac{e^{-x}}{2\sqrt{1+e^{-x}}}$ and $-\frac{e^{-x}\sqrt{1+e^{-x}}}{2(1+e^{-x})}$ are both correct answers. However, only the answer $-\frac{e^{-x}\sqrt{1+e^{-x}}}{2(1+e^{-x})}$ has a rational denominator. We multiplied $-\frac{e^{-x}}{2\sqrt{1+e^{-x}}}$ by $\frac{\sqrt{1+e^{-x}}}{\sqrt{1+e^{-x}}}$ in order to rationalize the denominator.

❽ $\frac{d}{dx}\frac{1-\sinh x}{\sinh x}$

Apply the quotient rule: $f(x) = 1 - \sinh x$, $g(x) = \sinh x$, $\frac{d}{dx}\left(\frac{f}{g}\right) = ?$

$$\frac{d}{dx}\left(\frac{f}{g}\right) = \frac{g\frac{df}{dx} - f\frac{dg}{dx}}{g^2} = \frac{\sinh x\left[\frac{d}{dx}(1-\sinh x)\right] - (1-\sinh x)\left(\frac{d}{dx}\sinh x\right)}{\sinh^2 x}$$

$$= \frac{\sinh x\,(0 - \cosh x) - (1-\sinh x)(\cosh x)}{\sinh^2 x}$$ Distribute: $-(a-b) = -a-(-b)$

$$= \frac{-\sinh x\cosh x - (1)\cosh x - (-\sinh x)\cosh x}{\sinh^2 x}$$ Note: $-(-\sinh x) = +\sinh x$

$$= \frac{-\sinh x\cosh x - \cosh x + \sinh x\cosh x}{\sinh^2 x} = \boxed{\frac{-\cosh x}{\sinh^2 x}}$$

Note: The answer may alternatively be expressed as $-\coth x\,\text{csch}\,x$ or $-\cosh x\,\text{csch}^2\,x$ because $\coth x = \frac{\cosh x}{\sinh x}$ and $\text{csch}\,x = \frac{1}{\sinh x}$.

Chapter 5, Part A

❶ $\frac{d}{dx}e^x \ln x$

Apply the product rule: $f(x) = e^x$, $g(x) = \ln x$, $\frac{d}{dx}(fg) = ?$

$$\frac{d}{dx}(fg) = g\frac{df}{dx} + f\frac{dg}{dx} = e^x\left(\frac{d}{dx}\ln x\right) + \ln x\left(\frac{d}{dx}e^x\right) = e^x\left(\frac{1}{x}\right) + \ln x\,(e^x)$$
$$= \frac{e^x}{x} + e^x \ln x = \boxed{e^x\left(\frac{1}{x} + \ln x\right)}$$

❷ $\frac{d}{dt} = \frac{\ln t}{t}$

Apply the quotient rule: $f(t) = \ln t$, $g(t) = t$, $\frac{d}{dt}\left(\frac{f}{g}\right) = ?$

$$\frac{d}{dt}\left(\frac{f}{g}\right) = \frac{g\frac{df}{dt} - f\frac{dg}{dt}}{g^2} = \frac{t\left(\frac{d}{dt}\ln t\right) - \ln t\left(\frac{d}{dt}t\right)}{t^2} = \frac{t\left(\frac{1}{t}\right) - \ln t\,(1)}{t^2} = \boxed{\frac{1 - \ln t}{t^2}}$$

❸ $\frac{d}{dx}\ln|\cos x|$

Apply the chain rule: $u = \cos x$, $f = \ln u$, $\frac{df}{dx} = ?$

$$\frac{df}{dx} = \frac{df}{du}\frac{du}{dx} = \left(\frac{d}{du}\ln u\right)\left(\frac{d}{dx}\cos x\right) = \left(\frac{1}{u}\right)(-\sin x) = \frac{-\sin x}{u} = -\frac{\sin x}{\cos x} = \boxed{-\tan x}$$

(Why put absolute values on $\ln|\cos x|$? The logarithm is only real when the argument is positive. The absolute values prevent cosine from being negative.)

❹ $\frac{d}{dx}\ln(\cosh x)$

Apply the chain rule: $u = \cosh x$, $f = \ln u$, $\frac{df}{dx} = ?$

$$\frac{df}{dx} = \frac{df}{du}\frac{du}{dx} = \left(\frac{d}{du}\ln u\right)\left(\frac{d}{dx}\cosh x\right) = \left(\frac{1}{u}\right)(\sinh x) = \frac{\sinh x}{u} = \frac{\sinh x}{\cosh x} = \boxed{\tanh x}$$

Note: The derivative of hyperbolic cosine, $\frac{d}{dx}(\cosh x) = \sinh x$, is positive, whereas the derivative of ordinary cosine, $\frac{d}{dx}\cos x = -\sin x$, is negative.

Chapter 5, Part B

❺ $\frac{d}{dx}\sqrt{\ln x}$

Apply the chain rule: $u = \ln x$, $f = \sqrt{u} = u^{1/2}$, $\frac{df}{dx} = ?$

$$\frac{df}{dx} = \frac{df}{du}\frac{du}{dx} = \left(\frac{d}{du}u^{1/2}\right)\left(\frac{d}{dx}\ln x\right) = \left(\frac{1}{2}u^{1/2-1}\right)\left(\frac{1}{x}\right) = \left(\frac{1}{2}u^{-1/2}\right)\left(\frac{1}{x}\right) = \left(\frac{1}{2u^{1/2}}\right)\left(\frac{1}{x}\right)$$

$$= \frac{1}{2xu^{1/2}} = \frac{1}{2x\sqrt{u}} = \frac{1}{2x\sqrt{\ln x}} = \frac{1}{2x\sqrt{\ln x}}\frac{\sqrt{\ln x}}{\sqrt{\ln x}} = \boxed{\frac{\sqrt{\ln x}}{2x\ln x}}$$

Notes:

- $\frac{1}{2} - 1 = \frac{1}{2} - \frac{2}{2} = \frac{1-2}{2} = -\frac{1}{2}$ (subtract fractions with a common denominator).
- $\frac{1}{x^{1/2}} = x^{-1/2}$ because $x^{-n} = \frac{1}{x^n}$.
- $\sqrt{\ln x}\sqrt{\ln x} = \ln x$ because $\sqrt{u}\sqrt{u} = \left(\sqrt{u}\right)^2 = u$.
- $\frac{1}{2x\sqrt{\ln x}}$ and $\frac{\sqrt{\ln x}}{2x\ln x}$ are both correct answers. However, only the answer $\frac{\sqrt{\ln x}}{2x\ln x}$ has a rational denominator. We multiplied $\frac{1}{2x\sqrt{\ln x}}$ by $\frac{\sqrt{\ln x}}{\sqrt{\ln x}}$ in order to rationalize the denominator.

❻ $\frac{d}{dx}\ln\sqrt{x}$

Method 1: Apply the chain rule: $u = \sqrt{x} = x^{1/2}$, $f = \ln u$, $\frac{df}{dx} = ?$

$$\frac{df}{dx} = \frac{df}{du}\frac{du}{dx} = \left(\frac{d}{du}\ln u\right)\left(\frac{d}{dx}x^{1/2}\right) = \left(\frac{1}{u}\right)\left(\frac{1}{2}x^{1/2-1}\right) = \left(\frac{1}{x^{1/2}}\right)\left(\frac{1}{2}x^{-1/2}\right)$$

$$= \left(\frac{1}{x^{1/2}}\right)\left(\frac{1}{2x^{1/2}}\right) = \frac{1}{2x^{1/2}x^{1/2}} = \frac{1}{2\sqrt{x}\sqrt{x}} = \boxed{\frac{1}{2x}}$$

Note that $\sqrt{x}\sqrt{x} = \left(\sqrt{x}\right)^2 = x$.

Method 2: Apply the identity $\ln x^a = a\ln x$ with $a = \frac{1}{2}$.

$$\frac{d}{dx}\ln\sqrt{x} = \frac{d}{dx}\ln x^{1/2} = \frac{d}{dx}\left(\frac{1}{2}\ln x\right) = \frac{1}{2}\frac{d}{dx}\ln x = \frac{1}{2}\left(\frac{1}{x}\right) = \boxed{\frac{1}{2x}}$$

❼ $\frac{d}{dt}\log_2 t$

Use the change of base formula with $b = 2$: $\log_2 t = \frac{\ln t}{\ln 2}$.

$$\frac{d}{dt}\log_2 t = \frac{d}{dt}\left(\frac{\ln t}{\ln 2}\right) = \frac{1}{\ln 2}\frac{d}{dt}\ln t = \frac{1}{\ln 2}\frac{1}{t} = \boxed{\frac{1}{t\ln 2}}$$

❽ $\frac{d}{dx}\frac{2^x}{x^2}$

Apply the quotient rule: $f(x) = 2^x$, $g(x) = x^2$, $\frac{d}{dx}\left(\frac{f}{g}\right) = ?$

$$\frac{d}{dx}\left(\frac{f}{g}\right) = \frac{g\frac{df}{dx} - f\frac{dg}{dx}}{g^2} = \frac{x^2\left(\frac{d}{dx}2^x\right) - 2^x\left(\frac{d}{dx}x^2\right)}{(x^2)^2}$$

Use the formula for the derivative of the power function with $b = 2$:

$$\frac{d}{dx}2^x = 2^x \ln 2$$

Substitute the above derivative into the previous equation.

$$\frac{d}{dx}\left(\frac{f}{g}\right) = \frac{x^2(2^x \ln 2) - 2^x(2x)}{(x^2)^2} = \frac{x^2 2^x \ln 2 - 2x2^x}{x^4}$$

$$= \boxed{\frac{2^x(x^2 \ln 2 - 2x)}{x^4}} = \boxed{\frac{2^x(x \ln 2 - 2)}{x^3}}$$

Notes:

- $(x^2)^2 = x^4$ according to $(x^a)^b = x^{ab}$.
- $x^2 2^x \ln 2 - 2x2^x = 2^x(x^2 \ln 2 - 2x)$. Factor out 2^x.
- In the last step, we divided the numerator and denominator both by x.
- An alternative answer is $\frac{2^x \ln 2}{x^2} - \frac{2^{x+1}}{x^3}$ because $\frac{x^2}{x^4} = \frac{1}{x^2}$ and $\frac{2^x 2x}{x^4} = \frac{2^x 2}{x^3} = \frac{2^{x+1}}{x^3}$ (since $2^x 2 = 2^x 2^1 = 2^{x+1}$ according to $x^m x^n = x^{m+n}$).

Chapter 6, Part A

❶ $\frac{dy}{dx} = \frac{d}{dx}(x^7 - 3x^5 + 5x^3 - 7x) = 7x^6 - 15x^4 + 15x^2 - 7$

$$\frac{d^2y}{dx^2} = \frac{d}{dx}\left(\frac{dy}{dx}\right) = \frac{d}{dx}(7x^6 - 15x^4 + 15x^2 - 7) = \boxed{42x^5 - 60x^3 + 30x}$$

❷ $\frac{dy}{d\theta} = \cos 3\theta$

Apply the chain rule with $u = 3\theta$ and $y = \cos u$:

$$\frac{dy}{d\theta} = \frac{dy}{du}\frac{du}{d\theta} = \left(\frac{d}{du}\cos u\right)\left(\frac{d}{d\theta}3\theta\right) = (-\sin u)(3) = -3\sin u = -3\sin 3\theta$$

Now take a second derivative:

$$\frac{d^2y}{d\theta^2} = \frac{d}{d\theta}\left(\frac{dy}{d\theta}\right) = \frac{d}{d\theta}(-3\sin 3\theta)$$

Apply the chain rule with $u = 3\theta$ and $g = -3\sin u$:

$$\frac{d^2y}{d\theta^2} = \frac{dg}{d\theta} = \frac{dg}{du}\frac{du}{d\theta} = \left[\frac{d}{du}(-3\sin u)\right]\left(\frac{d}{d\theta}3\theta\right)$$

$$= (-3\cos u)(3) = -9\cos u = \boxed{-9\cos 3\theta}$$

❸ $\frac{dy}{dx} = \frac{d}{dx}\ln x = \frac{1}{x} \quad \rightarrow \quad \frac{d^2y}{dx^2} = \frac{d}{dx}\left(\frac{dy}{dx}\right) = \frac{d}{dx}\left(\frac{1}{x}\right) = \frac{d}{dx}(x^{-1}) = -x^{-2} = \boxed{-\frac{1}{x^2}}$

❹ $\frac{dy}{dt} = \frac{d}{dt}e^{-3t^2}$

Apply the chain rule with $u = -3t^2$ and $y = e^u$:

$$\frac{dy}{dt} = \frac{dy}{du}\frac{du}{dt} = \left(\frac{d}{du}e^u\right)\left[\frac{d}{dt}(-3t^2)\right] = (e^u)(-6t) = -6te^u = -6te^{-3t^2}$$

Now take a second derivative:

$$\frac{d^2y}{dt^2} = \frac{d}{dt}\left(\frac{dy}{dt}\right) = \frac{d}{dt}\left(-6te^{-3t^2}\right)$$

Apply the product rule with $g(t) = -6t$ and $h(t) = e^{-3t^2}$.

$$\frac{d^2y}{dt^2} = \frac{d}{dt}(gh) = h\frac{dg}{dt} + g\frac{dh}{dt} = e^{-3t^2}\left[\frac{d}{dt}(-6t)\right] + (-6t)\left(\frac{d}{dt}e^{-3t^2}\right)$$

$$= e^{-3t^2}(-6) - 6t\left(-6te^{-3t^2}\right) = -6e^{-3t^2} + 36t^2e^{-3t^2} = (-6 + 36t^2)e^{-3t^2}$$

$$= (36t^2 - 6)e^{-3t^2} = 6(6t^2 - 1)e^{-3t^2} = \boxed{\frac{6(6t^2 - 1)}{e^{3t^2}}}$$

Notes:

- We used $\frac{d}{dt}e^{-3t^2} = -6te^{-3t^2}$ from $\frac{dy}{dt}$ when finding $\frac{d^2y}{dt^2}$.
- $-6e^{-3t^2} + 36t^2e^{-3t^2} = (-6 + 36t^2)e^{-3t^2}$. (Factor out e^{-3t^2}.)
- Note that $-6 + 36t^2 = 36t^2 - 6$.
- Each of the last five answers is a correct answer for this problem.

Chapter 6, Part B

❺ $\frac{dy}{d\theta} = \frac{d}{d\theta}\sin(\theta^2)$

Apply the chain rule with $u = \theta^2$ and $y = \sin u$:

$$\frac{dy}{d\theta} = \frac{dy}{du}\frac{du}{d\theta} = \left(\frac{d}{du}\sin u\right)\left(\frac{d}{d\theta}\theta^2\right) = (\cos u)(2\theta) = 2\theta\cos u = 2\theta\cos(\theta^2)$$

Now take a second derivative:

$$\frac{d^2y}{d\theta^2} = \frac{d}{d\theta}\left[\frac{dy}{d\theta}\right] = \frac{d}{d\theta}[2\theta\cos(\theta^2)]$$

Apply the product rule with $g(\theta) = 2\theta$ and $h(\theta) = \cos(\theta^2)$.

$$\frac{d^2y}{d\theta^2} = \frac{d}{d\theta}(gh) = h\frac{dg}{d\theta} + g\frac{dh}{d\theta} = \cos(\theta^2)\left(\frac{d}{d\theta}2\theta\right) + 2\theta\left[\frac{d}{d\theta}\cos(\theta^2)\right]$$

$$= \cos(\theta^2)\,(2) + 2\theta\left[\frac{d}{d\theta}\cos(\theta^2)\right] = 2\cos(\theta^2) + 2\theta\left[\frac{d}{d\theta}\cos(\theta^2)\right]$$

Apply the chain rule with $u = \theta^2$ and $k = \cos u$:

$$\frac{dk}{d\theta} = \frac{dk}{du}\frac{du}{d\theta} = \left(\frac{d}{du}\cos u\right)\left(\frac{d}{d\theta}\theta^2\right) = (-\sin u)(2\theta) = -2\theta\sin u = -2\theta\sin(\theta^2)$$

Substitute $\frac{d}{d\theta}\cos(\theta^2) = -2\theta\sin(\theta^2)$ into the equation for the second derivative:

$$\frac{d^2y}{d\theta^2} = 2\cos(\theta^2) + 2\theta\left[\frac{d}{d\theta}\cos(\theta^2)\right] = 2\cos(\theta^2) + 2\theta[-2\theta\sin(\theta^2)]$$

$$= \boxed{2\cos(\theta^2) - 4\theta^2\sin(\theta^2)}$$

Note that the angle is squared, not the sine function. Compare $\sin(\theta^2)$, which means to square the angle first and then take the sine, with $\sin^2\theta$, which instead means to take the sine first and then square the result.

❻ $\frac{dy}{d\theta} = \frac{d}{d\theta}\tan\theta = \sec^2\theta \quad \rightarrow \quad \frac{d^2y}{d\theta^2} = \frac{d}{d\theta}\left(\frac{dy}{d\theta}\right) = \frac{d}{d\theta}\sec^2\theta$

Apply the chain rule with $u = \sec\theta$ and $f = u^2$:

$$\frac{d^2y}{d\theta^2} = \frac{df}{d\theta} = \frac{df}{du}\frac{du}{d\theta} = \left(\frac{d}{du}u^2\right)\left(\frac{d}{d\theta}\sec\theta\right)$$

$$= (2u)(\sec\theta\tan\theta) = (2\sec\theta)(\sec\theta\tan\theta) = \boxed{2\sec^2\theta\tan\theta}$$

Note that an alternate answer is $\frac{2\sin\theta}{\cos^3\theta}$ because $\sec^2\theta = \frac{1}{\cos^2\theta}$ and $\tan\theta = \frac{\sin\theta}{\cos\theta}$.

❼ $\frac{dy}{dt} = \frac{d}{dt}(t \ln t)$

Apply the product rule with $f(t) = t$ and $g(t) = \ln t$.

$$\frac{dy}{dt} = \frac{d}{dt}(fg) = g\frac{df}{dt} + f\frac{dg}{dt} = \ln t\left(\frac{d}{dt}t\right) + t\left(\frac{d}{dt}\ln t\right) = (\ln t)(1) + t\left(\frac{1}{t}\right)$$
$$= \ln t + 1 = 1 + \ln t$$

Now take a second derivative:

$$\frac{d^2y}{dt^2} = \frac{d}{dt}\left(\frac{dy}{dt}\right) = \frac{d}{dt}(1 + \ln t) = \frac{d}{dt}1 + \frac{d}{dt}\ln t = 0 + \frac{1}{t} = \boxed{\frac{1}{t}}$$

❽ $\frac{dy}{dx} = \frac{d}{dx}\sqrt{x} = \frac{d}{dx}x^{\frac{1}{2}} = \frac{1}{2}x^{-\frac{1}{2}}$

$$\frac{d^2y}{dx^2} = \frac{d}{dx}\left(\frac{dy}{dx}\right) = \frac{d}{dx}\left(\frac{1}{2}x^{-\frac{1}{2}}\right) = \frac{1}{2}\frac{d}{dx}x^{-\frac{1}{2}} = \left(-\frac{1}{2}\right)\left(\frac{1}{2}\right)x^{-\frac{3}{2}}$$
$$= -\frac{1}{4}x^{-3/2} = -\frac{1}{4x^{3/2}} = -\frac{1}{4x\sqrt{x}} = -\frac{1}{4x\sqrt{x}}\frac{\sqrt{x}}{\sqrt{x}} = \boxed{-\frac{\sqrt{x}}{4x^2}}$$

Notes:

- $\frac{1}{2} - 1 = \frac{1}{2} - \frac{2}{2} = \frac{1-2}{2} = -\frac{1}{2}$ (subtract fractions with a common denominator).
- $-\frac{1}{2} - 1 = -\frac{1}{2} - \frac{2}{2} = \frac{-1-2}{2} = -\frac{3}{2}$.
- $\sqrt{x} = x^{1/2}$ and $\sqrt{x}\sqrt{x} = \left(\sqrt{x}\right)^2 = x$.
- $x^{-3/2} = \frac{1}{x^{3/2}}$ according to $x^{-n} = \frac{1}{x^n}$. Also, $x^{3/2} = x^1x^{1/2} = x\sqrt{x}$.
- There are multiple correct answers, but $-\frac{\sqrt{x}}{4x^2}$ has a rational denominator.

Chapter 6, Part C

❾ $\frac{dy}{dx} = \frac{d}{dx}\frac{1}{x} = \frac{d}{dx}x^{-1} = -x^{-2}$

$$\frac{d^2y}{dx^2} = \frac{d}{dx}\left(\frac{dy}{dx}\right) = \frac{d}{dx}(-x^{-2}) = (-2)(-1)x^{-3} = 2x^{-3} = \boxed{\frac{2}{x^3}}$$

Notes:

- $-1 - 1 = -2$ and $-2 - 1 = -3$.
- $(-2)(-1) = 2$.
- $x^{-1} = \frac{1}{x}$ and $x^{-3} = \frac{1}{x^3}$.
- $2x^{-3}$ and $\frac{2}{x^3}$ are both correct answers.

⑩ $\frac{dy}{d\theta} = \frac{d}{d\theta}\sec\theta = \sec\theta\tan\theta \quad \to \quad \frac{d^2y}{d\theta^2} = \frac{d}{d\theta}\left(\frac{dy}{d\theta}\right) = \frac{d}{d\theta}(\sec\theta\tan\theta)$

Apply the product rule with $f = \sec\theta$ and $g = \tan\theta$:

$$\frac{d^2y}{d\theta^2} = \frac{d}{d\theta}(fg) = g\frac{df}{d\theta} + f\frac{dg}{d\theta} = \tan\theta\left(\frac{d}{d\theta}\sec\theta\right) + \sec\theta\left(\frac{d}{d\theta}\tan\theta\right)$$

$$= \tan\theta(\sec\theta\tan\theta) + \sec\theta(\sec^2\theta) = \sec\theta\tan^2\theta + \sec^3\theta = \boxed{\sec\theta(\tan^2\theta + \sec^2\theta)}$$

An alternate answer is $\frac{\sin^2\theta+1}{\cos^3\theta}$ because $\sec\theta = \frac{1}{\cos\theta}$ and $\tan^2\theta = \frac{\sin^2\theta}{\cos^2\theta}$.

⑪ $\frac{dy}{dt} = \frac{d}{dt}\tan^{-1}t = \frac{1}{1+t^2} \quad \to \quad \frac{d^2y}{dt^2} = \frac{d}{dt}\left(\frac{dy}{dt}\right) = \frac{d}{dt}\left(\frac{1}{1+t^2}\right)$

The formula for the derivative of the inverse tangent was given in Chapter 3. Apply the chain rule with $u = 1 + t^2$ and $f = \frac{1}{u} = u^{-1}$.

$$\frac{d^2y}{dt^2} = \frac{df}{dt} = \frac{df}{du}\frac{du}{dt} = \left(\frac{d}{du}u^{-1}\right)\left[\frac{d}{dt}(1+t^2)\right] = (-u^{-2})(0+2t)$$

$$= -2tu^{-2} = -\frac{2t}{u^2} = \boxed{-\frac{2t}{(1+t^2)^2}}$$ An alternate answer is $= -\frac{2t}{1+2t^2+t^4}$.

⑫ $\frac{dy}{d\theta} = \frac{d}{d\theta}(\theta\sin\theta)$

Apply the product rule with $f = \theta$ and $g = \sin\theta$:

$$\frac{dy}{d\theta} = \frac{d}{d\theta}(fg) = g\frac{df}{d\theta} + f\frac{dg}{d\theta} = \sin\theta\left(\frac{d}{d\theta}\theta\right) + \theta\left(\frac{d}{d\theta}\sin\theta\right) = \sin\theta(1) + \theta\cos\theta$$

$$\frac{dy}{d\theta} = \sin\theta + \theta\cos\theta$$

Now take a second derivative:

$$\frac{d^2y}{d\theta^2} = \frac{d}{d\theta}\left[\frac{dy}{d\theta}\right] = \frac{d}{d\theta}(\sin\theta + \theta\cos\theta) = \cos\theta + \frac{d}{d\theta}(\theta\cos\theta)$$

Apply the product rule with $f = \theta$ and $h = \cos\theta$:

$$\frac{d}{d\theta}(\theta\cos\theta) = \frac{d}{d\theta}(fh) = h\frac{df}{d\theta} + f\frac{dh}{d\theta} = \cos\theta\left(\frac{d}{d\theta}\theta\right) + \theta\left(\frac{d}{d\theta}\cos\theta\right)$$

$$= \cos\theta(1) + \theta(-\sin\theta) = \cos\theta - \theta\sin\theta$$

Substitute $\frac{d}{d\theta}(\theta\cos\theta) = \cos\theta - \theta\sin\theta$ into the equation for $\frac{d^2y}{d\theta^2}$:

$$\frac{d^2y}{d\theta^2} = \cos\theta + \frac{d}{d\theta}(\theta\cos\theta) = \cos\theta + (\cos\theta - \theta\sin\theta) = \boxed{2\cos\theta - \theta\sin\theta}$$

Chapter 7

❶ $f(x) = 2x^4 - 8x^3 \quad , \quad 2 \leq x \leq 4$

Take a derivative of $f(x)$ with respect to x:

$$\frac{df}{dx} = \frac{d}{dx}(2x^4 - 8x^3) = 8x^3 - 24x^2$$

Set the first derivative equal to zero. Solve for x. Call these values x_c.

$$\frac{df}{dx} = 0 \quad \rightarrow \quad 8x_c^3 - 24x_c^2 = 0$$

Factor out $8x_c^2$:

$$8x_c^2(x_c - 3) = 0$$
$$8x_c^2 = 0 \quad \text{or} \quad x_c - 3 = 0$$
$$x_c = 0 \quad \text{or} \quad x_c = 3$$

Of these, only $x_c = 3$ lies in the specified interval, $2 \leq x \leq 4$.

Take a second derivative of $f(x)$ with respect to x:

$$\frac{d^2f}{dx^2} = \frac{d}{dx}\left(\frac{df}{dx}\right) = \frac{d}{dx}(8x^3 - 24x^2) = 24x^2 - 48x$$

Evaluate the second derivative at $x_c = 3$ (which we found previously).

$$\left.\frac{d^2f}{dx^2}\right|_{x=3} = 24(3)^2 - 48(3) = 216 - 144 = 72 \ \left(\begin{matrix}\text{relative} \\ \text{minimum}\end{matrix}\right)$$

There is a relative minimum at $x_c = 3$ (where the second derivative is positive).

Evaluate the function at $x_c = 3$, and the endpoints ($x = 2$ and $x = 4$).

$$f(3) = 2(3)^4 - 8(3)^3 = 162 - 216 = -54$$
$$f(2) = 2(2)^4 - 8(2)^3 = 32 - 64 = -32$$
$$f(4) = 2(4)^4 - 8(4)^3 = 512 - 512 = 0$$

Over the interval (2,4), the absolute extrema are:

- $f(x)$ has an absolute maximum value of $\boxed{0}$ (when $x = 4$).
- $f(x)$ has an absolute minimum value of $\boxed{-54}$ (when $x = 3$).

❷ $f(x) = \frac{4}{x^6} - \frac{3}{x^8}$, $\frac{1}{2} \leq x \leq 10$

Take a derivative of $f(x)$ with respect to x:

$$\frac{df}{dx} = \frac{d}{dx}\left(\frac{4}{x^6} - \frac{3}{x^8}\right) = \frac{d}{dx}(4x^{-6} - 3x^{-8}) = -24x^{-7} + 24x^{-9} = -\frac{24}{x^7} + \frac{24}{x^9}$$

Set the first derivative equal to zero. Solve for x. Call these values x_c.

$$\frac{df}{dx} = 0 \quad \to \quad -\frac{24}{x_c^7} + \frac{24}{x_c^9} = 0 \quad \to \quad \frac{24}{x_c^9} = \frac{24}{x_c^7} \quad \to \quad \frac{1}{x_c^9} = \frac{1}{x_c^7}$$

Multiply both sides of the equation by x_c^9:

$$1 = \frac{x_c^9}{x_c^7} \quad \to \quad 1 = x_c^2 \quad \to \quad \pm 1 = x_c$$

Note that $\frac{x_c^9}{x_c^7} = x_c^{9-7} = x_c^2$ because $\frac{x^m}{x^n} = x^{m-n}$. The reason that $x_c = -1$ and $x_c = 1$ both solve $x_c^2 = 1$ is that $(-1)^2 = 1$ and $1^2 = 1$. However, the problem specified the interval $\frac{1}{2} \leq x < 10$, and the only solution that satisfies this interval is $x_c = 1$.

Take a second derivative of $f(x)$ with respect to x:

$$\frac{d^2f}{dx^2} = \frac{d}{dx}\left(\frac{df}{dx}\right) = \frac{d}{dx}(-24x^{-7} + 24x^{-9}) = 168x^{-8} - 216x^{-10} = \frac{168}{x^8} - \frac{216}{x^{10}}$$

Evaluate the second derivative at $x_c = 1$ (which we found previously).

$$\left.\frac{d^2f}{dx^2}\right|_{x=1} = \frac{168}{1^8} - \frac{216}{1^{10}} = 168 - 216 = -48 \left(\begin{matrix}\text{relative} \\ \text{maximum}\end{matrix}\right)$$

There is a relative maximum at $x_c = 1$ (where the second derivative is negative).

Evaluate the function at $x_c = 1$, and the endpoints ($x = \frac{1}{2}$ and $x = 10$).

$$f(1) = \frac{4}{1^6} - \frac{3}{1^8} = 4 - 3 = 1$$

$$f\left(\frac{1}{2}\right) = \frac{4}{\left(\frac{1}{2}\right)^6} - \frac{3}{\left(\frac{1}{2}\right)^8} = 4(2)^6 - 3(2)^8 = 256 - 768 = -512$$

To divide by a fraction, multiply by its reciprocal: $\frac{4}{\left(\frac{1}{2}\right)^6} = 4\left(\frac{2}{1}\right)^6 = 4(2)^6 = 256$.

$$f(10) = \frac{4}{10^6} - \frac{3}{10^8} = 0.000004 - 0.00000003 = 0.00000397$$

Over the interval $\left(-\frac{1}{2}, 10\right)$, the absolute extrema are:

- $f(x)$ has an absolute maximum value of $\boxed{1}$ (when $x = 1$).
- $f(x)$ has an absolute minimum value of $\boxed{-512}$ (when $x = \frac{1}{2}$).

Chapter 8, Part A

❶ $$\lim_{x\to 2}\frac{x^2-3x+6}{x^2+3x-2}=\frac{2^2-3(2)+6}{2^2+3(2)-2}=\frac{4-6+6}{4+6-2}=\frac{4}{8}=\boxed{\frac{1}{2}}$$

Note that l'Hôpital's rule doesn't apply because neither the numerator nor the denominator approach 0 (or $\pm\infty$) in the specified limit.

❷ $$\lim_{x\to 0}\frac{\sqrt{144-x^2}}{\sqrt{9-x^2}}=\frac{\sqrt{144-0^2}}{\sqrt{9-0^2}}=\frac{\sqrt{144}}{\sqrt{9}}=\frac{12}{3}=\boxed{4}$$

Note that l'Hôpital's rule doesn't apply because neither the numerator nor the denominator approach 0 (or $\pm\infty$) in the specified limit.

❸ $$\lim_{x\to 1}e^x\ln x=e^1\ln(1)=e(0)=\boxed{0}$$

Note that $e^1=e$ and $\ln(1)=0$.

❹ $$\lim_{x\to\pi}x\cos x=\pi\cos\pi=\pi(-1)=\boxed{-\pi}$$

Note that $\cos\pi=\cos 180°=-1$.

❺ $$\lim_{x\to\infty}\frac{2x-8}{6x-4}=\lim_{x\to\infty}\frac{\frac{2x}{x}-\frac{8}{x}}{\frac{6x}{x}-\frac{4}{x}}=\lim_{x\to\infty}\frac{2-\frac{8}{x}}{6-\frac{4}{x}}=\frac{2-0}{6-0}=\frac{2}{6}=\boxed{\frac{1}{3}}$$

Notes:

- We divided the numerator and denominator each by x in the first step.
- $\frac{8}{x}$ and $\frac{4}{x}$ each approach 0 as x grows larger and larger.
- Calculator check: If you plug in $x=1000$, you will get $\frac{1992}{5996}\approx 0.33\approx\frac{1}{3}$.

❻ $$\lim_{x\to 4}\frac{x^2-16}{x-4}=\lim_{x\to 4}\frac{(x+4)(x-4)}{x-4}=\lim_{x\to 4}(x+4)=4+4=\boxed{8}$$

Notes:

- $(x+4)(x-4)=x^2+4x-4x-16=x^2-16$.
- We're exploring what happens to the function as x gets close to 4, but since x never quite reaches 4, we don't need to be worried about dividing by zero.
- Calculator check: If you plug in $x=4.1$, you will get $\frac{0.81}{0.1}=8.1\approx 8$.

❼ $$\lim_{x\to\infty}\frac{2x^2-3x}{4x^2+9x}=\lim_{x\to\infty}\frac{\frac{2x^2}{x^2}-\frac{3x}{x^2}}{\frac{4x^2}{x^2}+\frac{9x}{x^2}}=\lim_{x\to\infty}\frac{2-\frac{3}{x}}{4+\frac{9}{x}}=\frac{2-0}{4+0}=\frac{2}{4}=\boxed{\frac{1}{2}}$$

Notes:

- We divided the numerator and denominator each by x^2 in the first step.
- $\frac{3}{x}$ and $\frac{9}{x}$ each approach 0 as x grows larger and larger.
- Calculator check: If you plug in $x = 1000$, you will get $\frac{1{,}997{,}000}{4{,}009{,}000} \approx 0.5 = \frac{1}{2}$.

❽ $$\lim_{x\to 1}\frac{\ln x}{x-1}=\frac{\frac{d}{dx}\ln x\Big|_{x=1}}{\frac{d}{dx}(x-1)\Big|_{x=1}}=\frac{\frac{1}{x}\Big|_{x=1}}{1}=\frac{1/1}{1}=\frac{1}{1}=\boxed{1}$$

Notes:

- We applied l'Hôpital's rule since $\lim\limits_{x\to 1}\ln x = 0$ and $\lim\limits_{x\to 1}(x-1) = 0$.
- $\frac{d}{dx}\ln(x) = \frac{1}{x}$ and $\frac{d}{dx}(x-1) = \frac{d}{dx}x - \frac{d}{dx}1 = 1 - 0 = 1$.
- Note that $\frac{d}{dx}x = 1$ for any value of x.
- Calculator check: If you plug in $x = 1.01$, you will get $\frac{0.009950331}{0.01} \approx 0.995 \approx 1$.

Chapter 8, Part B

❾ $$\lim_{x\to 0}\frac{\tan x}{x}=\frac{\frac{d}{dx}\tan x\Big|_{x=0}}{\frac{d}{dx}x\Big|_{x=0}}=\frac{\sec^2 x|_{x=0}}{1}=\frac{\frac{1}{\cos^2 x}\Big|_{x=0}}{1}=\frac{\frac{1}{\cos^2 0}}{1}=\frac{1/1^2}{1}=\boxed{1}$$

Notes:

- We applied l'Hôpital's rule since $\lim\limits_{x\to 0}\tan x = 0$ and $\lim\limits_{x\to 0} x = 0$.
- $\frac{d}{dx}\tan x = \sec^2 x$ and $\frac{d}{dx}x = 1$.
- $\sec x = \frac{1}{\cos x}$, $\cos 0 = 1$, and $\sec 0 = 1$.
- Calculator check: If you plug in $x = 0.01$, you will get $\frac{0.01}{0.01} \approx 1$ (but first check that your calculator is in radians mode and not degrees mode).

⑩ $$\lim_{x\to\infty}\frac{2x^4+4x^2+6}{x^4+2x^2+3}=\lim_{x\to\infty}\frac{\frac{2x^4}{x^4}+\frac{4x^2}{x^4}+\frac{6}{x^4}}{\frac{x^4}{x^4}+\frac{2x^2}{x^4}+\frac{3}{x^4}}=\lim_{x\to\infty}\frac{2+\frac{4}{x^2}+\frac{6}{x^4}}{1+\frac{2}{x^2}+\frac{3}{x^4}}=\frac{2+0+0}{1+0+0}=\boxed{2}$$

Notes:

- We divided the numerator and denominator each by x^4 in the first step.
- $\frac{4}{x^2}$, $\frac{6}{x^4}$, $\frac{2}{x^2}$, and $\frac{3}{x^4}$ each approach 0 as x grows larger and larger.
- Calculator check: If you plug in $x = 100$, you will get $\frac{200{,}040{,}006}{100{,}020{,}003} \approx 2$.

⑪ $$\lim_{x\to\infty}\frac{x}{e^x}=\frac{\frac{d}{dx}x\Big|_{x\to\infty}}{\frac{d}{dx}e^x\Big|_{x\to\infty}}=\frac{1}{e^x|_{x\to\infty}}=\frac{1}{\lim\limits_{x\to\infty}e^x}=\boxed{0}$$

Notes:

- We applied l'Hôpital's rule since x and e^x both grow indefinitely as x gets larger and larger.
- $\frac{d}{dx}e^x = e^x$ and $\frac{d}{dx}x = 1$. Note that $\frac{d}{dx}x = 1$ for any value of x.
- As e^x grows larger and larger, $\frac{1}{e^x}$ gets closer and closer to zero.
- Calculator check: If you plug in $x = 20$, you will get $\frac{20}{485{,}165{,}195.4} \approx 0.000000041 \approx 0$.

⑫ $$\lim_{x\to 0}\frac{\sqrt{x+4}-2}{x}=\lim_{x\to 0}\frac{\sqrt{x+4}-2}{x}\frac{\sqrt{x+4}+2}{\sqrt{x+4}+2}=\lim_{x\to 0}\frac{x+4-4}{x\left(\sqrt{x+4}+2\right)}$$

$$=\lim_{x\to 0}\frac{x}{x\left(\sqrt{x+4}+2\right)}=\lim_{x\to 0}\frac{1}{\sqrt{x+4}+2}=\frac{1}{\sqrt{0+4}+2}=\frac{1}{\sqrt{4}+2}=\frac{1}{2+2}=\boxed{\frac{1}{4}}$$

Notes:

- $\left(\sqrt{x+4}-2\right)\left(\sqrt{x+4}+2\right)=\sqrt{x+4}\sqrt{x+4}+2\sqrt{x+4}-2\sqrt{x+4}-4$ $= x+4-4 = x$ for the same reason that $(a-b)(a+b) = a^2 - b^2$.
- Although x approaches 0, since it never quite reaches 0 in the limit, we don't need to worry about dividing by zero when we cancel $\frac{x}{x}$.
- If instead you apply l'Hôpital's rule, you will get the same answer. Note that $\frac{d}{dx}\sqrt{x+4} = \frac{1}{2\sqrt{x+4}}$, $\frac{d}{dx}x = 1$, and $\lim\limits_{x\to 0}\frac{1}{2\sqrt{x+4}} = \frac{1}{2\sqrt{4}} = \frac{1}{2(2)} = \frac{1}{4}$.
- Calculator check: If you plug in $x = 0.01$, you will get $\frac{0.002498439}{0.01} \approx 0.25 = \frac{1}{4}$.

⓭ $$\lim_{x\to\infty}\frac{8x^5-3x^3}{2x^6-9x^2}=\lim_{x\to\infty}\frac{\frac{8x^5}{x^6}-\frac{3x^3}{x^6}}{\frac{2x^6}{x^6}-\frac{9x^2}{x^6}}=\lim_{x\to\infty}\frac{\frac{8}{x}-\frac{3}{x^3}}{2-\frac{9}{x^4}}=\frac{0-0}{2-0}=\frac{0}{2}=\boxed{0}$$

Notes:

- We divided the numerator and denominator each by x^6 in the first step.
- $\frac{8}{x}$, $\frac{3}{x^3}$, and $\frac{9}{x^4}$ each approach 0 as x grows larger and larger.
- Conceptually, the denominator, which grows as x^6, dominates the numerator, which grows as x^5, causing the ratio to approach zero.
- Calculator check: If you plug in $x = 100$, you will get $\frac{7.9997\times10^{10}}{1.99999991\times10^{12}} \approx 0.04 \approx 0$.

⓮ $$\lim_{x\to0}\frac{e^x-e^{-x}}{x}=\frac{\frac{d}{dx}(e^x-e^{-x})\Big|_{x=0}}{\frac{d}{dx}x\Big|_{x=0}}=\frac{(e^x+e^{-x})|_{x=0}}{1}=\frac{1+1}{1}=\frac{2}{1}=\boxed{2}$$

Notes:

- We applied l'Hôpital's rule since $\lim\limits_{x\to0}(e^x-e^{-x})=0$ and $\lim\limits_{x\to0}x=0$.
- $\frac{d}{dx}e^x=e^x$, $\frac{d}{dx}e^{-x}=-e^{-x}$, and $\frac{d}{dx}x=1$.
- $\frac{d}{dx}(e^x-e^{-x})=\frac{d}{dx}e^x-\frac{d}{dx}e^{-x}=e^x-(-e^{-x})=e^x+e^{-x}$.
- $e^0=1$.
- Calculator check: If you plug in $x = 0.01$, you will get $\frac{0.020000333}{0.01}\approx 2$.

⓯ $$\lim_{x\to\frac{\pi}{2}}\frac{2x-\pi}{\cos x}=\frac{\frac{d}{dx}(2x-\pi)\Big|_{x=\frac{\pi}{2}}}{\frac{d}{dx}\cos x\Big|_{x=\frac{\pi}{2}}}=\frac{2}{-\sin x|_{x=\frac{\pi}{2}}}=\frac{2}{-\sin\frac{\pi}{2}}=\frac{2}{-1}=\boxed{-2}$$

Notes:

- We applied l'Hôpital's rule since $\lim\limits_{x\to\frac{\pi}{2}}(2x-\pi)=0$ and $\lim\limits_{x\to\frac{\pi}{2}}\cos x=0$.
- $\frac{d}{dx}\cos x=-\sin x$ and $\frac{d}{dx}(2x-\pi)=\frac{d}{dx}2x-\frac{d}{dx}\pi=2-0=2$.
- $\sin\frac{\pi}{2}=\sin 90°=1$ and $\cos\frac{\pi}{2}=\cos 90°=0$.
- Calculator check: If you plug in $x = 1.6$, you will get $\frac{0.058407346}{-0.029199522}\approx -2$ (but first check that your calculator is in radians mode and not degrees mode).

⓰ $$\lim_{x\to\infty}\frac{\ln x}{x} = \frac{\frac{d}{dx}\ln x\Big|_{x\to\infty}}{\frac{d}{dx}x\Big|_{x\to\infty}} = \frac{\frac{1}{x}\Big|_{x\to\infty}}{1} = \frac{0}{1} = \boxed{0}$$

Notes:

- We applied l'Hôpital's rule since $\ln x$ and x both grow indefinitely as x gets larger and larger.
- $\frac{d}{dx}\ln x = \frac{1}{x}$ and $\frac{d}{dx}x = 1$. Note that $\frac{d}{dx}x = 1$ for any value of x.
- Calculator check: If you plug in $x = 1{,}000{,}000$, you will get $\frac{13.81551056}{1{,}000{,}000} \approx 0.000013816 \approx 0$.

Chapter 9, Part A

❶ $$\int 35x^6\,dx = \frac{35x^{6+1}}{6+1} + c = \frac{35x^7}{7} + c = \boxed{5x^7 + c}$$

Check your answer:

$$\frac{d}{dx}(5x^7 + c) = 35x^6$$

❷ $$\int \frac{63}{t^8}\,dt = \int 63t^{-8}\,dt = \frac{63t^{-8+1}}{-8+1} + c = \frac{63t^{-7}}{-7} + c = -9t^{-7} + c = \boxed{-\frac{9}{t^7} + c}$$

Notes:

- $-8 + 1 = -7$ and $t^{-7} = \frac{1}{t^7}$.
- $-9t^{-7} + c$ and $-\frac{9}{t^7} + c$ are both correct answers.

Check your answer:

$$\frac{d}{dt}(-9t^{-7} + c) = 63t^{-8} = \frac{63}{t^8}$$

❸ $$\int 15x^{2/3}\,dx = \frac{15x^{2/3+1}}{\frac{2}{3}+1} + c = \frac{15x^{5/3}}{5/3} + c = \boxed{9x^{5/3} + c}$$

Note that $\frac{2}{3} + 1 = \frac{2}{3} + \frac{3}{3} = \frac{2+3}{3} = \frac{5}{3}$ and $15 \div \frac{5}{3} = 15 \times \frac{3}{5} = \frac{45}{5} = 9$.

Check your answer:

$$\frac{d}{dx}\left(9x^{5/3} + c\right) = \left(\frac{5}{3}\right)9x^{2/3} = \frac{45}{3}x^{2/3} = 15x^{2/3}$$

❹ $\int 48u^{-5}\,du = \frac{48u^{-5+1}}{-5+1} + c = \frac{48u^{-4}}{-4} + c = -12u^{-4} + c = \boxed{-\frac{12}{u^4} + c}$

Notes:

- $-5+1 = -4$ and $u^{-4} = \frac{1}{u^4}$.
- $-12u^{-4} + c$ and $-\frac{12}{u^4} + c$ are both correct answers.

Check your answer:

$$\frac{d}{du}(-12u^{-4} + c) = 48u^{-5}$$

Chapter 9, Part B

❺ $\int t\,dt = \int 1t^1\,dt = \frac{1t^{1+1}}{1+1} + c = \boxed{\frac{t^2}{2} + c}$

Note that $1t^1 = t$. Check your answer:

$$\frac{d}{dt}\left(\frac{t^2}{2} + c\right) = \frac{2t}{2} = t$$

❻ $\int \frac{dx}{\sqrt{x}} = \int \frac{dx}{x^{1/2}} = \int x^{-1/2}\,dx = \frac{x^{-1/2+1}}{-\frac{1}{2}+1} + c = \frac{x^{1/2}}{1/2} + c = 2x^{1/2} + c = \boxed{2\sqrt{x} + c}$

Notes:

- $\sqrt{x} = x^{1/2}$ and $x^{-1/2} = \frac{1}{x^{1/2}}$.
- $-\frac{1}{2} + 1 = -\frac{1}{2} + \frac{2}{2} = \frac{-1+2}{2} = \frac{1}{2}$ and $\frac{1}{1/2} = 1 \div \frac{1}{2} = 1 \times \frac{2}{1} = 2$.
- $2x^{1/2} + c$ and $2\sqrt{x} + c$ are both correct answers.

Check your answer:

$$\frac{d}{dx}(2x^{1/2} + c) = \left(\frac{1}{2}\right)2x^{-1/2} = x^{-1/2} = \frac{1}{x^{1/2}} = \frac{1}{\sqrt{x}}$$

❼ $\int 4\,du = \int 4u^0\,du = \frac{4u^{0+1}}{0+1} + c = \frac{4u^1}{1} + c = \boxed{4u + c}$

Note that $u^0 = 1$ and $u^1 = u$. Check your answer:

$$\frac{d}{du}(4u + c) = 4$$

❽ $$\int \frac{8dx}{x^{1/3}} = \int 8x^{-1/3}\,dx = \frac{8x^{-1/3+1}}{-\frac{1}{3}+1} + c = \frac{8x^{2/3}}{2/3} + c = \boxed{12x^{2/3} + c}$$

Notes:

- $x^{-1/3} = \frac{1}{x^{1/3}}$.
- $-\frac{1}{3} + 1 = -\frac{1}{3} + \frac{3}{3} = \frac{-1+3}{3} = \frac{2}{3}$ and $8 \div \frac{2}{3} = 8 \times \frac{3}{2} = \frac{24}{2} = 12$.

Check your answer:

$$\frac{d}{dx}\left(12x^{2/3} + c\right) = \left(\frac{2}{3}\right) 12x^{2/3-1} = \frac{24}{3}x^{-1/3} = 8x^{-1/3} = \frac{8}{x^{1/3}}$$

Chapter 9, Part C

❾ $$\int (x^2 - 3x + 4)\,dx = \int x^2\,dx - \int 3x\,dx + \int 4\,dx = \boxed{\frac{x^3}{3} - \frac{3x^2}{2} + 4x + c}$$

Check your answer:

$$\frac{d}{dx}\left(\frac{x^3}{3} - \frac{3x^2}{2} + 4x + c\right) = \frac{3x^2}{3} - \frac{6x}{2} + 4 = x^2 - 3x + 4$$

❿ $$\int \left(\frac{1}{u^2} - \frac{4}{u}\right) du = \int (u^{-2} - 4u^{-1})\,du = \int u^{-2}\,du - \int 4u^{-1}\,du$$

$$= \frac{u^{-2+1}}{-2+1} - 4\ln u + c = \frac{u^{-1}}{-1} - 4\ln u + c = -u^{-1} - 4\ln u + c = \boxed{-\frac{1}{u} - 4\ln u + c}$$

Notes:

- $\frac{1}{u^2} = u^{-2}$ and $\frac{1}{u} = u^{-1}$.
- $-2 + 1 = -1$.
- Recall that a power of $b = -1$ is a special case: $\int \frac{du}{u} = \int u^{-1}\,du = \ln u + c$ (whereas $\int au^b\,du = \frac{au^{b+1}}{b+1} + c$ when $b \neq -1$).

Check your answer:

$$\frac{d}{du}(-u^{-1} - 4\ln u + c) = u^{-2} - \frac{4}{u} = \frac{1}{u^2} - \frac{4}{u}$$

⓫ $$\int \left(10x^{3/2} + 6x^{1/2}\right) dx = \int 10x^{3/2}\, dx + \int 6x^{1/2}\, dx = \frac{10x^{5/2}}{5/2} + \frac{6x^{3/2}}{3/2} + c$$
$$= \frac{20x^{5/2}}{5} + \frac{12x^{3/2}}{3} + c = \boxed{4x^{5/2} + 4x^{3/2} + c}$$

Notes:

- $\frac{3}{2} + 1 = \frac{3}{2} + \frac{2}{2} = \frac{3+2}{2} = \frac{5}{2}$ and $\frac{1}{2} + 1 = \frac{1}{1} + \frac{2}{2} = \frac{1+2}{2} = \frac{3}{2}$.
- $10 \div \frac{5}{2} = 10 \times \frac{2}{5} = \frac{20}{5} = 4$ and $6 \div \frac{3}{2} = 6 \times \frac{2}{3} = \frac{12}{3} = 4$.

Check your answer:
$$\frac{d}{dx}\left(4x^{5/2} + 4x^{3/2} + c\right) = \left(\frac{5}{2}\right)(4)x^{3/2} + \left(\frac{3}{2}\right)(4)x^{1/2} = \frac{20}{2}x^{3/2} + \frac{12}{2}x^{1/2}$$
$$= 10x^{3/2} + 6x^{1/2}$$

⓬ $$\int \left(\sqrt{t} + \frac{1}{\sqrt{t}}\right) dt = \int \left(t^{1/2} + \frac{1}{t^{1/2}}\right) dt = \int \left(t^{1/2} + t^{-1/2}\right) dt$$
$$= \int t^{1/2}\, dt + \int t^{-1/2}\, dt = \frac{t^{1/2+1}}{\frac{1}{2}+1} + \frac{t^{-1/2+1}}{-\frac{1}{2}+1} + c$$
$$= \frac{t^{3/2}}{3/2} + \frac{t^{1/2}}{1/2} + c = \boxed{\frac{2t^{3/2}}{3} + 2t^{1/2} + c} = \frac{2t\sqrt{t}}{3} + 2\sqrt{t} + c$$

Notes:

- $\frac{1}{2} + 1 = \frac{1}{1} + \frac{2}{2} = \frac{1+2}{2} = \frac{3}{2}$ and $-\frac{1}{2} + 1 = -\frac{1}{1} + \frac{2}{2} = \frac{-1+2}{2} = \frac{1}{2}$.
- $1 \div \frac{3}{2} = 1 \times \frac{2}{3} = \frac{2}{3}$ and $1 \div \frac{1}{2} = 1 \times \frac{2}{1} = 2$. Note that $t^{3/2} = t^1 t^{1/2} = t\sqrt{t}$.

Check your answer:
$$\frac{d}{dt}\left(\frac{2t^{3/2}}{3} + 2t^{1/2} + c\right) = \left(\frac{3}{2}\right)\left(\frac{2}{3}\right)t^{1/2} + \left(\frac{1}{2}\right)(2)t^{-1/2} = t^{1/2} + t^{-1/2} = \sqrt{t} + \frac{1}{\sqrt{t}}$$

Chapter 10, Part A

❶ $$\int_{x=1}^{2} 8x^3\, dx = \left[\frac{8x^{3+1}}{3+1}\right]_{x=1}^{2} = \left[\frac{8x^4}{4}\right]_{x=1}^{2} = [2x^4]_{x=1}^{2} = 2(2)^4 - 2(1)^4$$
$$= 2(16) - 2(1) = 32 - 2 = \boxed{30}$$

Check your antiderivative (by taking a derivative):
$$\frac{d}{dx}2x^4 = 8x^3$$

❷ $$\int_{x=3}^{6} \frac{dx}{x^3} = \int_{x=3}^{6} x^{-3}\,dx = \left[\frac{x^{-3+1}}{-3+1}\right]_{x=3}^{6} = \left[\frac{x^{-2}}{-2}\right]_{x=3}^{6} = \left[-\frac{x^{-2}}{2}\right]_{x=3}^{6} = \left[-\frac{1}{2x^2}\right]_{x=3}^{6}$$

$$= -\frac{1}{2(6^2)} - \left[-\frac{1}{2(3^2)}\right] = -\frac{1}{72} + \frac{1}{18} = -\frac{1}{72} + \frac{4}{72} = \frac{-1+4}{72} = \frac{3}{72} = \boxed{\frac{1}{24}} \approx \boxed{0.0417}$$

Check your antiderivative (by taking a derivative):

$$\frac{d}{dx}\left(-\frac{x^{-2}}{2}\right) = (-2)\left(-\frac{1}{2}\right)x^{-3} = x^{-3} = \frac{1}{x^3}$$

❸ $$\int_{t=4}^{9} \frac{dt}{\sqrt{t}} = \int_{t=4}^{9} \frac{dt}{t^{1/2}} = \int_{t=4}^{9} t^{-1/2}\,dt = \left[\frac{t^{-1/2+1}}{-\frac{1}{2}+1}\right]_{t=4}^{9} = \left[\frac{t^{1/2}}{1/2}\right]_{t=4}^{9} = \left[2t^{1/2}\right]_{t=4}^{9}$$

$$= \left[2\sqrt{t}\right]_{t=4}^{9} = 2\sqrt{9} - 2\sqrt{4} = 2(3) - 2(2) = 6 - 4 = \boxed{2}$$

Check your antiderivative (by taking a derivative):

$$\frac{d}{dt}\left(2t^{1/2}\right) = \left(\frac{1}{2}\right)(2)t^{-1/2} = t^{-1/2} = \frac{1}{t^{1/2}} = \frac{1}{\sqrt{t}}$$

❹ $$\int_{x=-2}^{2} (x^3 - 6x^2 + 4x - 8)\,dx = \left[\frac{x^4}{4} - \frac{6x^3}{3} + \frac{4x^2}{2} - 8x\right]_{x=-2}^{2}$$

$$= \left[\frac{x^4}{4} - 2x^3 + 2x^2 - 8x\right]_{x=-2}^{2}$$

$$= \frac{2^4}{4} - 2(2)^3 + 2(2)^2 - 8(2) - \left[\frac{(-2)^4}{4} - 2(-2)^3 + 2(-2)^2 - 8(-2)\right]$$

$$= \frac{16}{4} - 2(8) + 2(4) - 16 - \left[\frac{16}{4} - 2(-8) + 2(4) + 16\right]$$

$$= 4 - 16 + 8 - 16 - (4 + 16 + 8 + 16) = -20 - (44) = -20 - 44 = \boxed{-64}$$

Check your antiderivative (by taking a derivative):

$$\frac{d}{dx}\left(\frac{x^4}{4} - \frac{6x^3}{3} + \frac{4x^2}{2} - 8x\right) = \frac{4x^3}{4} - \frac{(3)(6)x^2}{3} + \frac{(2)(4)x}{2} - 8 = x^3 - 6x^2 + 4x - 8$$

Chapter 10, Part B

❺ $$\int_{x=0}^{5} \frac{x^4}{5}\,dx = \frac{1}{5}\int_{x=0}^{5} x^4\,dx = \frac{1}{5}\left[\frac{x^{4+1}}{4+1}\right]_{x=0}^{5} = \frac{1}{5}\left[\frac{x^5}{5}\right]_{x=0}^{5} = \frac{1}{5}\left(\frac{5^5}{5} - \frac{0^5}{5}\right) = \frac{1}{5}\left(\frac{5^5}{5} - 0\right)$$

$$= \frac{1}{5}\left(\frac{5^5}{5}\right) = \frac{5^5}{5(5)} = \frac{5^5}{5^2} = 5^{5-2} = 5^3 = \boxed{125}$$

Check your antiderivative (by taking a derivative):

$$\frac{d}{dx}\left(\frac{1}{5}\frac{x^5}{5}\right) = \frac{5x^4}{25} = \frac{x^4}{5}$$

❻ $$\int_{t=-3}^{6} (t^2 - 2)\,dt = \left[\frac{t^3}{3} - 2t\right]_{t=-3}^{6} = \frac{6^3}{3} - 2(6) - \left[\frac{(-3)^3}{3} - 2(-3)\right]$$

$$= \frac{216}{3} - 12 - \left(-\frac{27}{3} + 6\right) = 72 - 12 - (-9 + 6) = 60 - (-3) = 60 + 3 = \boxed{63}$$

Check your antiderivative (by taking a derivative):

$$\frac{d}{dt}\left(\frac{t^3}{3} - 2t\right) = \frac{3t^2}{3} - 2 = t^2 - 2$$

❼ $$\int_{x=1}^{4} \sqrt{x}\,dx = \int_{x=1}^{4} x^{1/2}\,dx = \left[\frac{x^{1/2+1}}{\frac{1}{2}+1}\right]_{x=1}^{4} = \left[\frac{x^{3/2}}{3/2}\right]_{x=1}^{4} = \left[\frac{2x^{3/2}}{3}\right]_{x=1}^{4}$$

$$= \frac{2(4)^{3/2}}{3} - \frac{2(1)^{3/2}}{3} = \frac{2(8)}{3} - \frac{2(1)}{3} = \frac{16}{3} - \frac{2}{3} = \frac{16-2}{3} = \boxed{\frac{14}{3}} \approx \boxed{4.667}$$

Note that $(4)^{3/2} = (4^3)^{1/2} = 64^{1/2} = \sqrt{64} = 8$, or use a calculator to verify that 4^(3/2) = 8.

Check your antiderivative (by taking a derivative):

$$\frac{d}{dx}\left(\frac{2x^{3/2}}{3}\right) = \left(\frac{3}{2}\right)\left(\frac{2}{3}\right)x^{1/2} = x^{1/2} = \sqrt{x}$$

❽ $\displaystyle\int_{x=1}^{e} \frac{dx}{x} = [\ln x]_{x=1}^{e} = \ln e - \ln 1 = 1 - 0 = \boxed{1}$

Notes:

- Recall from Chapter 9 that $\int \frac{dx}{x} = \int x^{-1}\,dx = \ln x + c$ (but, as usual, the constant of integration doesn't matter in a definite integral because it would cancel out during the subtraction after plugging in the limits).
- Recall from Chapter 5 that $\ln e = 0$ and $\ln 1 = 0$.
- We first encountered Euler's number, e, in Chapter 4.

Check your antiderivative (it may help to review Chapter 5):

$$\frac{d}{dx}\ln x = \frac{1}{x}$$

Chapter 11

❶ $\displaystyle\int_{\theta=0}^{\pi} \sin\theta\,d\theta = [-\cos\theta]_{\theta=0}^{\pi} = -\cos\pi - (-\cos 0) = -(-1) + 1 = 1 + 1 = \boxed{2}$

Notes:

- π rad $= 180°$ and $\cos(\pi) = \cos(180°) = -1$.
- $\cos 0 = 1$.
- $-\cos\pi = -(-1) = 1$ and $-(-\cos 0) = \cos 0 = 1$.

Check your antiderivative (it may help to review Chapter 3):

$$\frac{d}{d\theta}(-\cos\theta) = -\frac{d}{d\theta}\cos\theta = -(-\sin\theta) = \sin\theta$$

❷ $\displaystyle\int_{\theta=0}^{\pi/3} \tan\theta\,d\theta = [\ln|\sec\theta|]_{\theta=0}^{\frac{\pi}{3}} = \ln\left|\sec\frac{\pi}{3}\right| - \ln|\sec 0| = \ln 2 - \ln 1$

$$\ln 2 - 0 = \boxed{\ln 2} \approx \boxed{0.693}$$

Notes:

- $\frac{\pi}{3}$ rad $= \frac{\pi}{3}\frac{180°}{\pi} = 60°$, $\cos\left(\frac{\pi}{3}\right) = \cos(60°) = \frac{1}{2}$, and $\sec\left(\frac{\pi}{3}\right) = \frac{1}{\cos\left(\frac{\pi}{3}\right)} = \frac{1}{1/2} = 2$.
- $\cos 0 = 1$ and $\sec 0 = \frac{1}{\cos 0} = \frac{1}{1} = 1$.
- $\ln 1 = 0$. Use a calculator to determine that $\ln 2 \approx 0.693147181$.

❸ $$\int_{\theta=-\frac{\pi}{6}}^{\pi/6} \sec\theta\, d\theta = [\ln|\sec\theta + \tan\theta|]_{\theta=-\frac{\pi}{6}}^{\frac{\pi}{6}}$$

$$= \ln\left|\sec\frac{\pi}{6} + \tan\frac{\pi}{6}\right| - \ln\left|\sec\left(-\frac{\pi}{6}\right) + \tan\left(-\frac{\pi}{6}\right)\right|$$

$$= \ln\left|\frac{2}{\sqrt{3}} + \frac{1}{\sqrt{3}}\right| - \ln\left|\frac{2}{\sqrt{3}} - \frac{1}{\sqrt{3}}\right| = \ln\left|\frac{2+1}{\sqrt{3}}\right| - \ln\left|\frac{2-1}{\sqrt{3}}\right| = \ln\left|\frac{3}{\sqrt{3}}\right| - \ln\left|\frac{1}{\sqrt{3}}\right|$$

$$= \ln\left|\frac{3}{\sqrt{3}} \div \frac{1}{\sqrt{3}}\right| = \ln\left|\frac{3}{\sqrt{3}} \times \frac{\sqrt{3}}{1}\right| = \boxed{\ln 3} \approx \boxed{1.099}$$

Notes:

- $\frac{\pi}{6}$ rad $= \frac{\pi}{6}\frac{180°}{\pi} = 30°$.
- $\cos\left(\frac{\pi}{6}\right) = \cos(30°) = \frac{\sqrt{3}}{2}$ and $\cos\left(-\frac{\pi}{6}\right) = \cos(-30°) = \frac{\sqrt{3}}{2}$.
- $\sec\left(\frac{\pi}{6}\right) = \frac{1}{\cos\left(\frac{\pi}{6}\right)} = \frac{1}{\sqrt{3}/2} = \frac{2}{\sqrt{3}}$ and $\sec\left(-\frac{\pi}{6}\right) = \frac{1}{\cos\left(-\frac{\pi}{6}\right)} = \frac{1}{\sqrt{3}/2} = \frac{2}{\sqrt{3}}$.
- $\tan\left(\frac{\pi}{6}\right) = \tan(30°) = \frac{1}{\sqrt{3}}$ and $\tan\left(-\frac{\pi}{6}\right) = \tan(-30°) = -\frac{1}{\sqrt{3}}$.
- Are you wondering if $\tan(30°) = \frac{\sqrt{3}}{3}$ instead of $\tan(30°) = \frac{1}{\sqrt{3}}$? You shouldn't be wondering this because both are the same. That's because $\frac{1}{\sqrt{3}} = \frac{\sqrt{3}}{3}$, as you can see by rationalizing the denominator: $\frac{1}{\sqrt{3}} = \frac{1}{\sqrt{3}}\frac{\sqrt{3}}{\sqrt{3}} = \frac{\sqrt{3}}{3}$.
- Note that $\sec\left(-\frac{\pi}{6}\right)$ is positive whereas $\tan\left(-\frac{\pi}{6}\right)$ is negative. That's because secant is an even function (symmetric about the vertical axis) whereas tangent is an odd function (with an anti-symmetric graph).
- How does $\ln\left|\frac{3}{\sqrt{3}}\right| - \ln\left|\frac{1}{\sqrt{3}}\right|$ equal $\ln\left|\frac{3}{\sqrt{3}} \div \frac{1}{\sqrt{3}}\right|$? The way to see this is to apply the property of logarithms that $\ln\left(\frac{p}{q}\right) = \ln p - \ln q$, setting $p = \frac{3}{\sqrt{3}}$ and $q = \frac{1}{\sqrt{3}}$.
- $\frac{3}{\sqrt{3}} \div \frac{1}{\sqrt{3}} = \frac{3}{\sqrt{3}} \times \frac{\sqrt{3}}{1} = 3$. Recall that the way to divide two fractions is to multiply by the reciprocal of the second fraction. The reciprocal of $\frac{1}{\sqrt{3}}$ is $\frac{\sqrt{3}}{1}$.
- Use a calculator to determine that $\ln 3 \approx 1.098612289$.

❹ $$\int_{\theta=0}^{\pi/3} (\theta + \cos\theta)\,d\theta = \int_{\theta=0}^{\pi/3} \theta\,d\theta + \int_{\theta=0}^{\pi/3} \cos\theta\,d\theta = \left[\frac{\theta^2}{2}\right]_{\theta=0}^{\frac{\pi}{3}} + [\sin\theta]_{\theta=0}^{\frac{\pi}{3}}$$

$$= \frac{\left(\frac{\pi}{3}\right)^2}{2} - \frac{0^2}{2} + \sin\left(\frac{\pi}{3}\right) - \sin 0 = \frac{\left(\frac{\pi^2}{9}\right)}{2} - 0 + \frac{\sqrt{3}}{2} - 0 = \boxed{\frac{\pi^2}{18} + \frac{\sqrt{3}}{2}}$$

$$\approx 0.548311356 + 0.866025404 \approx \boxed{1.414}$$

Notes:

- The given integral is like $\int (y_1 + y_2)\,dx = \int y_1\,dx + \int y_2\,dx$, with $y_1 = \theta$ and $y_2 = \cos\theta$.
- The first integral has the same form as $\int x\,dx = \frac{x^2}{2}$, with θ in place of x.
- $\frac{\pi}{3}$ rad $= \frac{\pi}{3}\frac{180°}{\pi} = 60°$ and $\sin\left(\frac{\pi}{3}\right) = \sin(60°) = \frac{\sqrt{3}}{2}$.
- $\sin 0 = 0$.
- $\left(\frac{\pi}{3}\right)^2 = \frac{\pi^2}{9}$ because $\left(\frac{x}{y}\right)^2 = \frac{x^2}{y^2}$.
- $\frac{\left(\frac{\pi^2}{9}\right)}{2} = \frac{\pi^2}{9} \div 2 = \frac{\pi^2}{9} \div \frac{2}{1} = \frac{\pi^2}{9} \times \frac{1}{2} = \frac{\pi^2}{18}$. To divide by the fraction $\frac{2}{1}$, multiply by its reciprocal, which is $\frac{1}{2}$.
- We used a calculator to get the numerical values at the end of the solution.

Check your antiderivative (it may help to review Chapter 3):

$$\frac{d}{d\theta}\left(\frac{\theta^2}{2} + \sin\theta\right) = \frac{d}{d\theta}\frac{\theta^2}{2} + \frac{d}{d\theta}\sin\theta = \frac{2\theta}{2} + \cos\theta = \theta + \cos\theta$$

Chapter 12, Part A

❶ $$\int_{x=0}^{\infty} e^{-x}\,dx = \left[\frac{e^{-x}}{-1}\right]_{x=0}^{\infty} = [-e^{-x}]_{x=0}^{\infty} = \left[-\frac{1}{e^x}\right]_{x=0}^{\infty} = -\lim_{L\to\infty}\frac{1}{e^L} - \left(-\frac{1}{e^0}\right)$$

$$= -0 + \frac{1}{1} = \boxed{1}$$

This is an 'improper' integral: Imagine that the upper limit is $x = L$, such that the integral is $\int_{x=0}^{L} e^{-x}\,dx$. Integration gives $\left[-\frac{1}{e^x}\right]_{x=0}^{L} = -\frac{1}{e^L} - \left(-\frac{1}{e^0}\right)$. In the limit that L grows to ∞, the expression $\frac{1}{e^L}$ approaches zero.

❷ $$\int_{x=1}^{e} \ln x\, dx = [x \ln x - x]_{x=1}^{e} = e \ln e - e - (1 \ln 1 - 1) = e(1) - e - [1(0) - 1]$$

$$= e - e - (0 - 1) = e - e - (-1) = e - e + 1 = 0 + 1 = \boxed{1}$$

Notes:

- $\int \ln x\, dx = x \ln x - x + c$ (but as usual, we may ignore the constant of integration when performing a definite integral, since it would cancel out during the subtraction after plugging in the limits).
- $\ln e = 1$.
- $\ln 1 = 0$.

Check your antiderivative (it may help to review the product rule from Chapter 2 in addition to Chapter 5):

$$\frac{d}{dx}(x \ln x - x) = \frac{d}{dx}(x \ln x) - \frac{d}{dx}x = \left(x \frac{d}{dx} \ln x + \ln x \frac{d}{dx}\right) - \frac{d}{dx}x$$

$$= x\left(\frac{1}{x}\right) + \ln x\,(1) - (1) = 1 + \ln x - 1 = \ln x$$

❸ $$\int_{t=1}^{2} \cosh t\, dt = [\sinh t]_{t=1}^{2} = \boxed{\sinh 2 - \sinh 1} = \frac{e^2 - e^{-2}}{2} - \frac{e^1 - e^{-1}}{2}$$

$$\approx 3.626860408 - 1.175201194 \approx \boxed{2.452}$$

Notes:

- $\sinh x = \frac{e^x - e^{-x}}{2}$. This is a hyperbolic function (**not** the sine function from trig).
- We used a calculator to get the numerical values at the end of the solution.

❹ $$\int_{x=3}^{4} 2^x\, dx = \left[\frac{2^x}{\ln 2}\right]_{x=3}^{4} = \frac{2^4}{\ln 2} - \frac{2^3}{\ln 2} = \frac{2^4 - 2^3}{\ln 2} = \frac{16 - 8}{\ln 2} = \boxed{\frac{8}{\ln 2}} \approx \boxed{11.542}$$

Note that we used a calculator to get the numerical value in the last step.

Check your antiderivative (it may help to review Chapter 5):

$$\frac{d}{dx}\left(\frac{2^x}{\ln 2}\right) = \frac{1}{\ln 2}\frac{d}{dx}(2^x) = \frac{1}{\ln 2}(2^x \ln 2) = 2^x$$

Chapter 13, Part A

❶ $\int \frac{dx}{2x+3}$

$$u = 2x + 3 \quad , \quad du = 2\,dx \quad , \quad dx = \frac{du}{2}$$

$$\int \frac{dx}{2x+3} = \int \frac{1}{2x+3} dx = \int \frac{1}{u}\left(\frac{du}{2}\right) = \int \frac{du}{2u} = \frac{1}{2}\int \frac{du}{u}$$

$$= \frac{1}{2}\ln u + c = \boxed{\frac{1}{2}\ln(2x+3) + c}$$

Recall from Chapter 9 that an exponent of $b = -1$ is a special case of integrals of the form $\int au^b\,du$: In this case, $\int \frac{du}{u} = \int u^{-1}\,du = \ln u + c$.

❷ $\int\limits_{\theta=0}^{\pi/9} 6\sin(3\theta)\,d\theta$

$$u = 3\theta \quad , \quad du = 3\,d\theta \quad , \quad d\theta = \frac{du}{3}$$

$$u_1 = u(0) = 3(0) = 0 \quad , \quad u_2 = u\left(\frac{\pi}{9}\right) = 3\left(\frac{\pi}{9}\right) = \frac{\pi}{3}$$

$$\int\limits_{\theta=0}^{\pi/9} 6\sin(3\theta)\,d\theta = \int\limits_{u=0}^{\pi/3} 6\sin u \frac{du}{3} = \int\limits_{u=0}^{\pi/3} 2\sin u\,du = 2[-\cos u]_{u=0}^{\frac{\pi}{3}} = -2[\cos u]_{u=0}^{\frac{\pi}{3}}$$

$$= -2\left(\cos\frac{\pi}{3} - \cos 0\right) = -2\left(\frac{1}{2} - 1\right) = -2\left(-\frac{1}{2}\right) = \boxed{1}$$

Notes:

- $\frac{\pi}{3}$ rad $= \frac{\pi}{3}\frac{180°}{\pi} = 60°$.
- $\cos\left(\frac{\pi}{3}\right) = \cos(60°) = \frac{1}{2}$.
- $\cos 0 = 1$.
- $\int \sin\theta\,d\theta = -\cos\theta + c$ (but as usual, we may ignore the constant of integration when performing a definite integral, since it would cancel out during the subtraction after plugging in the limits).

Chapter 13, Part B

❸ $\int \frac{dx}{\sqrt{x-1}}$

$$u = x - 1 \quad , \quad du = dx \quad , \quad dx = du$$

$$\int \frac{dx}{\sqrt{x-1}} = \int \frac{du}{\sqrt{u}} = \int \frac{du}{u^{1/2}} = \int u^{-1/2}\, du = \frac{u^{-1/2+1}}{-\frac{1}{2}+1} + c = \frac{u^{1/2}}{1/2} + c = 2u^{1/2} + c$$

$$= 2\sqrt{u} + c = \boxed{2\sqrt{x-1} + c}$$

Notes:

- $\frac{du}{dx} = \frac{d}{dx}(x-1) = \frac{d}{dx}x - \frac{d}{dx}1 = 1 - 0 = 1$. Therefore, $du = 1\,dx = dx$.
- $\sqrt{u} = u^{1/2}$ and $u^{-1/2} = \frac{1}{u^{1/2}}$.
- $-\frac{1}{2} + 1 = -\frac{1}{2} + \frac{2}{2} = \frac{-1+2}{2} = \frac{1}{2}$ and $\frac{1}{1/2} = 1 \div \frac{1}{2} = 1 \times \frac{2}{1} = \frac{2}{1} = 2$.

❹ $\int_{t=0}^{\infty} t\, e^{-t^2}\, dt$

$$u = t^2 \quad , \quad du = 2t\, dt \quad , \quad dt = \frac{du}{2t}$$

$$u_1 = u(0) = 0^2 = 0 \quad , \quad u_2 = u(\infty) = (\infty)^2 = \infty$$

$$\int_{t=0}^{\infty} t\, e^{-t^2}\, dt = \int_{u=0}^{\infty} t\, e^{-u}\left(\frac{du}{2t}\right) = \int_{u=0}^{\infty} \frac{t\, e^{-u}}{2t} du = \frac{1}{2}\int_{u=0}^{\infty} e^{-u}\, du = \frac{1}{2}[-e^{-u}]_{u=0}^{\infty}$$

$$= -\frac{1}{2}[e^{-u}]_{u=0}^{\infty} = -\frac{1}{2}\left(\lim_{L\to\infty} e^{-L} - e^{-0}\right) = -\frac{1}{2}\left(0 - \frac{1}{e^0}\right) - \frac{1}{2}\left(0 - \frac{1}{1}\right) = -\frac{1}{2}(-1) = \boxed{\frac{1}{2}}$$

This is an 'improper' integral: Imagine that the upper limit is L, such that the integral is $\frac{1}{2}\int_{u=0}^{L} e^{-u}\, du$. Integration gives $\frac{1}{2}[-e^{-u}]_{u=0}^{L} = \frac{1}{2}[-e^{-L} - (-e^0)]$. In the limit that L grows to ∞, the expression $e^{-L} = \frac{1}{e^L}$ approaches zero, such that $\frac{1}{2}[-e^{-L} - (-e^0)]$ approaches $\frac{1}{2}[-0 - (-1)] = \frac{1}{2}(-0 + 1) = \frac{1}{2}(1) = \frac{1}{2}$.

Chapter 13, Part C

❺ $\int \frac{8x^3}{x^4 - 2} dx$

$$u = x^4 - 2 \quad , \quad du = 4x^3\, dx \quad , \quad dx = \frac{du}{4x^3}$$

$$\int \frac{8x^3}{x^4 - 2} dx = \int \frac{8x^3}{u}\frac{du}{4x^3} = \int \frac{2}{u} du = 2\int \frac{du}{u} = 2\ln u + c = \boxed{2\ln(x^4 - 2) + c}$$

Notes:

- $\frac{du}{dx} = \frac{d}{dx}(x^4 - 2) = \frac{d}{dx}x^4 - \frac{d}{dx}2 = 4x^3 - 0 = 4x^3$. Therefore, $du = 4x^3\, dx$.
- Recall from Chapter 9 that an exponent of $b = -1$ is a special case of integrals of the form $\int au^b\, du$: In this case, $\int \frac{du}{u} = \int u^{-1}\, du = \ln u + c$.

❻ $\int_{\theta=0}^{\frac{\sqrt{\pi}}{2}} \theta \cos(\theta^2)\, d\theta$

$$u = \theta^2 \quad , \quad du = 2\theta\, d\theta \quad , \quad d\theta = \frac{du}{2\theta}$$

$$u_1 = u(0) = 0^2 = 0 \quad , \quad u_2 = u\left(\frac{\sqrt{\pi}}{2}\right) = \left(\frac{\sqrt{\pi}}{2}\right)^2 = \frac{\pi}{4}$$

$$\int_{\theta=0}^{\frac{\sqrt{\pi}}{2}} \theta \cos(\theta^2)\, d\theta = \int_{u=0}^{\pi/4} \theta \cos u \frac{du}{2\theta} = \frac{1}{2}\int_{u=0}^{\pi/4} \cos u\, du = \frac{1}{2}[\sin u]_{u=0}^{\frac{\pi}{4}}$$

$$= \frac{1}{2}\left(\sin\frac{\pi}{4} - \sin 0\right) = \frac{1}{2}\left(\frac{\sqrt{2}}{2} - 0\right) = \frac{1}{2}\left(\frac{\sqrt{2}}{2}\right) = \boxed{\frac{\sqrt{2}}{4}}$$

Notes:

- $\frac{\pi}{4}\text{ rad} = \frac{\pi}{4}\frac{180°}{\pi} = 45°$.
- $\sin\left(\frac{\pi}{4}\right) = \sin(45°) = \frac{\sqrt{2}}{2}$.
- $\sin 0 = 0$.
- $\int \cos\theta\, d\theta = \sin\theta + c$ (but as usual, we may ignore the constant of integration when performing a definite integral, since it would cancel out during the subtraction after plugging in the limits).

Chapter 13, Part D

❼ $\int \frac{12x^2}{(x^3+16)^2} dx$

$$u = x^3 + 16 \quad , \quad du = 3x^2\, dx \quad , \quad dx = \frac{du}{3x^2}$$

$$\int \frac{12x^2}{(x^3+16)^2} dx = \int \frac{12x^2}{u^2}\frac{du}{3x^2} = \int \frac{4}{u^2} du = 4\int \frac{du}{u^2} = 4\int u^{-2} du = \frac{4u^{-2+1}}{-2+1} + c$$

$$= \frac{4u^{-1}}{-1} + c = -4u^{-1} + c = -\frac{4}{u} + c = \boxed{-\frac{4}{x^3+16} + c}$$

Notes:

- $\frac{du}{dx} = \frac{d}{dx}(x^3+16) = \frac{d}{dx}x^3 + \frac{d}{dx}16 = 3x^2 + 0 = 3x^2$. Therefore, $du = 3x^2\, dx$.
- $-2+1 = -1$, $u^{-2} = \frac{1}{u^2}$, and $u^{-1} = \frac{1}{u}$.

❽ $\int_{\theta=0}^{\pi^2/4} \frac{\sin(\sqrt{\theta})}{\sqrt{\theta}} d\theta$

$$u = \sqrt{\theta} = \theta^{1/2} \quad , \quad du = \frac{1}{2}\theta^{-1/2}\, d\theta = \frac{d\theta}{2\theta^{1/2}} = \frac{d\theta}{2\sqrt{\theta}} \quad , \quad d\theta = 2\sqrt{\theta}\, du$$

$$u_1 = u(0) = \sqrt{0} = 0 \quad , \quad u_2 = u\left(\frac{\pi^2}{4}\right) = \sqrt{\frac{\pi^2}{4}} = \frac{\pi}{2}$$

$$\int_{\theta=0}^{\pi^2/4} \frac{\sin(\sqrt{\theta})}{\sqrt{\theta}} d\theta = \int_{u=0}^{\pi/2} \frac{\sin u}{\sqrt{\theta}}(2\sqrt{\theta}\, du) = 2\int_{u=0}^{\pi/2} \sin u\, du = 2[-\cos u]_{u=0}^{\frac{\pi}{2}}$$

$$= -2[\cos u]_{u=0}^{\frac{\pi}{2}} = -2\left(\cos\frac{\pi}{2} - \cos 0\right) = -2(0-1) = -2(-1) = \boxed{2}$$

Notes:

- $\frac{\pi}{2}$ rad $= \frac{\pi}{2}\frac{180°}{\pi} = 90°$.
- $\cos\left(\frac{\pi}{2}\right) = \cos(90°) = 0$.
- $\cos 0 = 1$.
- $\int \sin\theta\, d\theta = -\cos\theta + c$ (but as usual, we may ignore the constant of integration when performing a definite integral, since it would cancel out during the subtraction after plugging in the limits).

Chapter 13, Part E

❾ $\int \sqrt{1-\sqrt{x}}\,dx$

$$u = 1 - \sqrt{x} = 1 - x^{1/2} \quad , \quad du = -\frac{1}{2}x^{-1/2}\,dx = -\frac{dx}{2x^{1/2}} = -\frac{dx}{2\sqrt{x}} \quad , \quad dx = -2\sqrt{x}\,du$$

$$\int \sqrt{1-\sqrt{x}}\,dx = \int \sqrt{u}\left(-2\sqrt{x}\,du\right) = -2\int \sqrt{u}\sqrt{x}\,du$$

Add $\sqrt{x}$ to both sides of the equation $u = 1 - \sqrt{x}$ to get $u + \sqrt{x} = 1$. Subtract u from both sides to get $\sqrt{x} = 1 - u$. Therefore, we may replace $\sqrt{x}$ with $1 - u$.

$$-2\int \sqrt{u}\sqrt{x}\,du = -2\int \sqrt{u}(1-u)\,du = -2\int u^{1/2}(1-u^1)\,du$$

$$= -2\int u^{1/2}\,du - 2\int u^{1/2}(-u^1)\,du = -2\int u^{1/2}\,du + 2\int u^{1/2}u^1\,du$$

$$= -2\int u^{1/2}\,du + 2\int u^{3/2}\,du = -\frac{2u^{1/2+1}}{\frac{1}{2}+1} + \frac{2u^{3/2+1}}{\frac{3}{2}+1} + c$$

$$= -\frac{2u^{3/2}}{3/2} + \frac{2u^{5/2}}{5/2} + c = -\frac{4u^{3/2}}{3} + \frac{4u^{5/2}}{5} + c$$

$$= \frac{4u^{5/2}}{5} - \frac{4u^{3/2}}{3} + c = \boxed{\frac{4\left(1-\sqrt{x}\right)^{5/2}}{5} - \frac{4\left(1-\sqrt{x}\right)^{3/2}}{3} + c}$$

Notes:

- $\frac{du}{dx} = \frac{d}{dx}\left(1 - x^{1/2}\right) = \frac{d}{dx}1 - \frac{d}{dx}x^{1/2} = 0 - \frac{1}{2}x^{-1/2} = -\frac{1}{2}x^{-1/2} = -\frac{dx}{2x^{1/2}} = -\frac{dx}{2\sqrt{x}}$. Therefore, $du = -\frac{dx}{2\sqrt{x}}$, such that $dx = -2\sqrt{x}\,du$.
- $\sqrt{u}(1-u) = \sqrt{u} - u\sqrt{u} = u^{1/2} - uu^{1/2} = u^{1/2} - u^1u^{1/2} = u^{1/2} - u^{3/2}$ because $x^m x^n = x^{m+n}$.
- Note that $u = u^1$ and $\sqrt{u} = u^{1/2}$.
- $\frac{1}{2} + 1 = \frac{1}{2} + \frac{2}{2} = \frac{1+2}{2} = \frac{3}{2}$ and $\frac{3}{2} + 1 = \frac{3}{2} + \frac{2}{2} = \frac{3+2}{2} = \frac{5}{2}$.
- $\frac{2}{3/2} = 2 \div \frac{3}{2} = 2 \times \frac{2}{3} = \frac{4}{3}$ and $\frac{2}{5/2} = 2 \div \frac{5}{2} = 2 \times \frac{2}{5} = \frac{4}{5}$.
- Going from $-\frac{4u^{3/2}}{3} + \frac{4u^{5/2}}{5} + c$ to $\frac{4u^{5/2}}{5} - \frac{4u^{3/2}}{3} + c$, we applied the identity $-a + b = b - a$.

⑩ $\int_{x=0}^{4} \sqrt{3x+4}\,dx$

$$u = 3x + 4 \quad , \quad du = 3\,dx \quad , \quad dx = \frac{du}{3}$$

$$u_1 = u(0) = 3(0) + 4 = 4 \quad , \quad u_2 = u(4) = 3(4) + 4 = 12 + 4 = 16$$

$$\int_{x=0}^{4} \sqrt{3x+4}\,dx = \int_{u=4}^{16} \sqrt{u}\,\frac{du}{3} = \frac{1}{3}\int_{u=4}^{16} \sqrt{u}\,du = \frac{1}{3}\int_{u=4}^{16} u^{1/2}\,du = \frac{1}{3}\left[\frac{u^{1/2+1}}{\frac{1}{2}+1}\right]_{u=4}^{u=16}$$

$$= \frac{1}{3}\left[\frac{u^{3/2}}{3/2}\right]_{u=4}^{u=16} = \frac{1}{3}\left[\frac{2u^{3/2}}{3}\right]_{u=4}^{u=16} = \frac{2}{9}\left[u^{3/2}\right]_{u=4}^{u=16} = \frac{2}{9}\left(16^{3/2} - 4^{3/2}\right)$$

$$= \frac{2}{9}(64 - 8) = \frac{2}{9}(56) = \boxed{\frac{112}{9}} \approx \boxed{12.444}$$

Notes:

- $\frac{d}{dx}(3x+4) = \frac{d}{dx}3x + \frac{d}{dx}4 = 3 + 0 = 3$. Therefore, $du = 3\,dx$.
- $u^{1/2} = \sqrt{u}$.
- $\frac{1}{2} + 1 = \frac{1}{2} + \frac{2}{2} = \frac{1+2}{2} = \frac{3}{2}$ and $\frac{1}{3/2} = 1 \div \frac{3}{2} = 1 \times \frac{2}{3} = \frac{2}{3}$.
- Note that $(16)^{3/2} = \left(16^{1/2}\right)^3 = 4^3 = 64$ and $(4)^{3/2} = \left(4^{1/2}\right)^3 = 2^3 = 8$ because $x^{mn} = (x^m)^n = (x^n)^m$. Alternatively, use a calculator to verify that 16^(3/2) = 64 and 4^(3/2) = 8.

Chapter 14, Part A

❶ $\int \frac{dx}{\sqrt{16 - x^2}} = \int \frac{dx}{\sqrt{4^2 - x^2}}$

$$x = a\sin\theta = 4\sin\theta \quad , \quad dx = a\cos\theta\,d\theta = 4\cos\theta\,d\theta$$

$$\int \frac{dx}{\sqrt{4^2 - x^2}} = \int \frac{4\cos\theta\,d\theta}{\sqrt{4^2 - (4\sin\theta)^2}} = \int \frac{4\cos\theta\,d\theta}{\sqrt{4^2 - 4^2\sin^2\theta}} = \int \frac{4\cos\theta\,d\theta}{\sqrt{4^2(1 - \sin^2\theta)}}$$

$$= \int \frac{4\cos\theta\,d\theta}{\sqrt{4^2\cos^2\theta}} = \int \frac{4\cos\theta\,d\theta}{4\cos\theta} = \int d\theta = \theta + c = \boxed{\sin^{-1}\left(\frac{x}{4}\right) + c}$$

Notes:

- $4^2 - 4^2\sin^2\theta = 4^2(1 - \sin^2\theta) = 4^2\cos^2\theta$ because $\sin^2\theta + \cos^2\theta = 1$.
- In the last step, $\theta = \sin^{-1}\left(\frac{x}{4}\right)$ follows from $x = 4\sin\theta \to \frac{x}{4} = \sin\theta$.

❷ $\int_{\theta=0}^{\pi/3} \cos^4\theta \sin\theta \, d\theta$

$$u = \cos\theta \quad , \quad du = -\sin\theta \, d\theta$$

$$u_1 = u(0) = \cos 0 = 1 \quad , \quad u_2 = u\left(\frac{\pi}{3}\right) = \cos\left(\frac{\pi}{3}\right) = \cos(60°) = \frac{1}{2}$$

$$\int_{\theta=0}^{\pi/3} \cos^4\theta \sin\theta \, d\theta = \int_{u=1}^{1/2} u^4 (-du) = -\int_{u=1}^{\frac{1}{2}} u^4 \, du = -\left[\frac{u^5}{5}\right]_{u=1}^{\frac{1}{2}} = -\frac{1}{5}[u^5]_{u=1}^{\frac{1}{2}}$$

$$= -\frac{1}{5}\left(\frac{1}{2^5} - 1^5\right) = -\frac{1}{5}\left(\frac{1}{32} - 1\right) = -\frac{1}{5}\left(\frac{1}{32} - \frac{32}{32}\right) = -\frac{1}{5}\left(\frac{1-32}{32}\right)$$

$$= -\frac{1}{5}\left(-\frac{31}{32}\right) = \boxed{\frac{31}{160}} \approx \boxed{0.194}$$

Notes:

- $\frac{\pi}{3}\text{rad} = \frac{\pi}{3}\frac{180°}{\pi} = 60°$.
- $\left(\frac{1}{2}\right)^5 = \frac{1}{2^5} = \frac{1}{32}$.

Chapter 14, Part B

❸ $\int \frac{dx}{x^2 + 25} = \int \frac{dx}{x^2 + 5^2}$

$$x = a\tan\theta = 5\tan\theta \quad , \quad dx = a\sec^2\theta \, d\theta = 5\sec^2\theta \, d\theta$$

$$\int \frac{dx}{x^2 + 5^2} = \int \frac{5\sec^2\theta \, d\theta}{(5\tan\theta)^2 + 5^2} = \int \frac{5\sec^2\theta \, d\theta}{5^2\tan^2\theta + 5^2} = \int \frac{5\sec^2\theta \, d\theta}{5^2(\tan^2\theta + 1)}$$

$$= \int \frac{5\sec^2\theta \, d\theta}{5^2\sec^2\theta} = \int \frac{d\theta}{5} = \frac{1}{5}\int d\theta = \frac{\theta}{5} + c = \boxed{\frac{1}{5}\tan^{-1}\left(\frac{x}{5}\right) + c}$$

Notes:

- $5^2\tan^2\theta + 5^2 = 5^2(\tan^2\theta + 1) = 5^2\sec^2\theta$ because $\tan^2\theta + 1 = \sec^2\theta$.
- In the last step, $\theta = \tan^{-1}\left(\frac{x}{5}\right)$ follows from $x = 5\tan\theta \rightarrow \frac{x}{5} = \tan\theta$.

❹ $$\int_{\theta=0}^{\pi/4} \tan^3\theta\, d\theta = \int_{\theta=0}^{\pi/4} \tan^2\theta \tan\theta\, d\theta = \int_{\theta=0}^{\pi/4} (\sec^2\theta - 1)\tan\theta\, d\theta$$

$$= \int_{\theta=0}^{\pi/4} \sec^2\theta \tan\theta\, d\theta - \int_{\theta=0}^{\pi/4} \tan\theta\, d\theta$$

Note that $\tan^2\theta = \sec^2\theta - 1$ follows from $\tan^2\theta + 1 = \sec^2\theta$. Distribute to see that $(\sec^2\theta - 1)\tan\theta = \sec^2\theta\tan\theta - \tan\theta$.

$$u = \tan\theta \quad , \quad du = \sec^2\theta\, d\theta$$

$$u_1 = u(0) = \tan 0 = 1 \quad , \quad u_2 = u\left(\frac{\pi}{4}\right) = \tan\left(\frac{\pi}{4}\right) = \tan(45°) = 1$$

$$\int_{\theta=0}^{\pi/4} \sec^2\theta \tan\theta\, d\theta - \int_{\theta=0}^{\pi/4} \tan\theta\, d\theta = \int_{u=0}^{1} u\, du - \int_{\theta=0}^{\pi/4} \tan\theta\, d\theta$$

$$= \left[\frac{u^2}{2}\right]_{u=0}^{1} - [\ln|\sec\theta|]_{\theta=0}^{\frac{\pi}{4}} = \frac{1^2}{2} - \frac{0^2}{2} - \left(\ln\left|\sec\frac{\pi}{4}\right| - \ln|\sec 0|\right)$$

$$= \frac{1}{2} - 0 - \left(\ln\sqrt{2} - \ln 1\right) = \frac{1}{2} - \left(\ln\sqrt{2} - 0\right)$$

$$= \boxed{\frac{1}{2} - \ln\sqrt{2}} = \boxed{\frac{1}{2} - \ln 2^{1/2}} = \boxed{\frac{1}{2} - \frac{1}{2}\ln 2} = \boxed{\frac{1}{2}(1 - \ln 2)} \approx \boxed{0.153}$$

Notes:

- On the last line, we applied the rule $\ln x^a = a\ln x$ to write $\ln 2^{1/2} = \frac{1}{2}\ln 2$.
- $\frac{\pi}{4}\text{rad} = \frac{\pi}{4}\frac{180°}{\pi} = 45°$.
- $\sec\frac{\pi}{4} = \sec 45° = \frac{1}{\cos 45°} = \frac{2}{\sqrt{2}} = \sqrt{2}$ and $\sec 0 = \frac{1}{\cos 0} = \frac{1}{1} = 1$. Note that $\frac{2}{\sqrt{2}} = \sqrt{2}$ because $\sqrt{2}\sqrt{2} = 2$.
- $\int \tan\theta\, d\theta = \ln|\sec\theta| + c$ (but as usual, we may ignore the constant of integration when performing a definite integral, since it would cancel out during the subtraction after plugging in the limits).
- Recall from Chapter 5 that $\ln 1 = 0$.
- We used a calculator to get the numerical value at the end of the solution.
- An alternative way to solve this problem is to write $\tan^3\theta = \frac{\sin^3\theta}{\cos^3\theta}$, rewrite $\sin^3\theta = \sin^2\theta\sin\theta = (1 - \cos^2\theta)\sin\theta$, separate the integral into two integrals, and make the substitution $u = \sin\theta$ and $du = \cos\theta\, d\theta$.

Chapter 14, Part C

❺ $\int \sin^4\theta\, d\theta = \int \sin^2\theta \sin^2\theta\, d\theta = \int (1-\cos^2\theta)\sin^2\theta\, d\theta$

$$= \int (\sin^2\theta - \cos^2\theta\sin^2\theta)\, d\theta = \int (\sin^2\theta - \sin^2\theta\cos^2\theta)\, d\theta$$

Note that $\sin^2\theta = 1 - \cos^2\theta$ follows from $\sin^2\theta + \cos^2\theta = 1$. Distribute to see that $(1-\cos^2\theta)\sin^2\theta = \sin^2\theta - \cos^2\theta\sin^2\theta$.

$$\int (\sin^2\theta - \sin^2\theta\cos^2\theta)\, d\theta = \int [\sin^2\theta - (\sin\theta\cos\theta)^2]\, d\theta$$

$$= \int \left[\sin^2\theta - \left(\frac{\sin 2\theta}{2}\right)^2\right] d\theta = \int \left(\sin^2\theta - \frac{\sin^2 2\theta}{4}\right) d\theta$$

$$= \int \left(\frac{1-\cos 2\theta}{2} - \frac{1-\cos 4\theta}{8}\right) d\theta = \int \left(\frac{4-4\cos 2\theta}{8} - \frac{1-\cos 4\theta}{8}\right) d\theta$$

$$= \frac{1}{8}\int [4 - 4\cos 2\theta - (1-\cos 4\theta)]\, d\theta = \frac{1}{8}\int (4 - 4\cos 2\theta - 1 + \cos 4\theta)\, d\theta$$

$$= \frac{1}{8}\int (3 - 4\cos 2\theta + \cos 4\theta)\, d\theta = \frac{1}{8}\int 3\, d\theta - \frac{1}{8}\int 4\cos 2\theta\, d\theta + \frac{1}{8}\int \cos 4\theta\, d\theta$$

Notes:

- $\frac{\sin^2 2\theta}{4} = \sin^2\theta\cos^2\theta$ because $\sin 2\theta = 2\sin\theta\cos\theta \rightarrow \frac{\sin 2\theta}{2} = \sin\theta\cos\theta$.
- $\sin^2\theta = \frac{1-\cos 2\theta}{2}$ and $\sin^2 2\theta = \frac{1-\cos 4\theta}{2}$ (double the angles on both sides).

$$u = 2\theta \quad , \quad du = 2\, d\theta \quad , \quad d\theta = \frac{du}{2} \quad \text{(second integral)}$$

$$v = 4\theta \quad , \quad dv = 4\, d\theta \quad , \quad d\theta = \frac{dv}{4} \quad \text{(third integral)}$$

$$\frac{1}{8}\int 3\, d\theta - \frac{1}{8}\int 4\cos 2\theta\, d\theta + \frac{1}{8}\int \cos 4\theta\, d\theta = \frac{3}{8}\int d\theta - \frac{4}{8}\int \cos u \frac{du}{2} + \frac{1}{8}\int \cos v \frac{dv}{4}$$

$$= \frac{3}{8}\int d\theta - \frac{1}{2}\int \cos u \frac{du}{2} + \frac{1}{8}\int \cos v \frac{dv}{4} = \frac{3}{8}\int d\theta - \frac{1}{4}\int \cos u\, du + \frac{1}{32}\int \cos v\, dv$$

$$= \frac{3}{8}\theta - \frac{1}{4}\sin u + \frac{1}{32}\sin v + c = \boxed{\frac{3}{8}\theta - \frac{1}{4}\sin 2\theta + \frac{1}{32}\sin 4\theta + c}$$

❻ $$\int_{x=0}^{1} \frac{dx}{(x^2+1)^2} = \int_{x=0}^{1} \frac{dx}{(x^2+1^2)^2}$$

$$x = a\tan\theta = 1\tan\theta = \tan\theta \quad , \quad dx = a\sec^2\theta\, d\theta = 1\sec^2\theta\, d\theta = \sec^2\theta\, d\theta$$

Since $x = \tan\theta$, it follows that $\theta = \tan^{-1}(x)$.

$$\theta_1 = \tan^{-1}(0) = 0 \quad , \quad \theta_1 = \tan^{-1}(1) = 45° = \frac{\pi}{4}\text{rad}$$

$$\int_{x=0}^{1} \frac{dx}{(x^2+1)^2} = \int_{\theta=0}^{\pi/4} \frac{\sec^2\theta\, d\theta}{(\tan^2\theta+1)^2} = \int_{\theta=0}^{\pi/4} \frac{\sec^2\theta\, d\theta}{(\sec^2\theta)^2} = \int_{\theta=0}^{\pi/4} \frac{\sec^2\theta\, d\theta}{\sec^4\theta}$$

$$= \int_{\theta=0}^{\pi/4} \frac{d\theta}{\sec^2\theta} = \int_{\theta=0}^{\pi/4} \cos^2\theta\, d\theta = \int_{\theta=0}^{\pi/4} \frac{1+\cos 2\theta}{2} d\theta$$

$$= \int_{\theta=0}^{\pi/4} \frac{1}{2} d\theta + \int_{\theta=0}^{\pi/4} \frac{\cos 2\theta}{2} d\theta = \frac{1}{2}\int_{\theta=0}^{\pi/4} d\theta + \frac{1}{2}\int_{\theta=0}^{\pi/4} \cos 2\theta\, d\theta$$

$$u = 2\theta \quad , \quad du = 2\,d\theta \quad , \quad d\theta = \frac{du}{2}$$

$$u_1 = 2(0) = 0 \quad , \quad u_2 = 2\left(\frac{\pi}{4}\right) = \frac{\pi}{2}$$

$$\frac{1}{2}[\theta]_{\theta=0}^{\frac{\pi}{4}} + \frac{1}{2}\int_{u=0}^{\pi/2} \cos u \frac{du}{2} = \frac{1}{2}\left(\frac{\pi}{4} - 0\right) + \frac{1}{4}\int_{u=0}^{\pi/2} \cos u\, du$$

$$= \frac{1}{2}\left(\frac{\pi}{4}\right) + \frac{1}{4}[\sin u]_{u=0}^{\pi/2} = \frac{\pi}{8} + \frac{1}{4}\left[\sin\left(\frac{\pi}{2}\right) - \sin 0\right]$$

$$= \frac{\pi}{8} + \frac{1}{4}(1-0) = \boxed{\frac{\pi}{8} + \frac{1}{4}} = \frac{\pi}{8} + \frac{2}{8} = \boxed{\frac{\pi+2}{8}} \approx \boxed{0.643}$$

Notes:

- $\tan^2\theta + 1 = \sec^2\theta$.
- $\sec\theta = \frac{1}{\cos\theta}$ such that $\frac{1}{\sec\theta} = \cos\theta$. Recall the identity $\cos^2\theta = \frac{1+\cos 2\theta}{2}$.
- $\frac{\pi}{4}\text{rad} = \frac{\pi}{4}\frac{180°}{\pi} = 45°$ and $\frac{\pi}{2}\text{rad} = \frac{\pi}{2}\frac{180°}{\pi} = 90°$.
- $\sin\frac{\pi}{2} = \sin 90° = 1$ and $\sin 0 = 0$.
- Note carefully that the upper limit of the θ integrals is $\frac{\pi}{4}$, whereas the upper limit of the u integral is $\frac{\pi}{2}$ because $u = 2\theta$.
- We used a calculator to get the numerical value at the end of the solution.

Chapter 14, Part D

❼ $\int \frac{dx}{(9-x^2)^{3/2}} = \int \frac{dx}{(3^2-x^2)^{3/2}}$

$$x = a\sin\theta = 3\sin\theta \quad , \quad dx = a\cos\theta\, d\theta = 3\cos\theta\, d\theta$$

$$\int \frac{dx}{(3^2-x^2)^{3/2}} = \int \frac{3\cos\theta\, d\theta}{[3^2-(3\sin\theta)^2]^{3/2}} = 3\int \frac{\cos\theta\, d\theta}{(3^2-3^2\sin^2\theta)^{3/2}}$$

$$= 3\int \frac{\cos\theta\, d\theta}{[3^2(1-\sin^2\theta)]^{3/2}} = 3\int \frac{\cos\theta\, d\theta}{(3^2\cos^2\theta)^{3/2}} = 3\int \frac{\cos\theta\, d\theta}{3^3\cos^3\theta} = \frac{3}{3^3}\int \frac{\cos\theta\, d\theta}{\cos^3\theta}$$

$$= \frac{1}{3^2}\int \frac{d\theta}{\cos^2\theta} = \frac{1}{9}\int \sec^2\theta\, d\theta = \frac{1}{9}\tan\theta + c = \frac{\sin\theta}{9\cos\theta} + c = \frac{\sin\theta}{9\sqrt{1-\sin^2\theta}} + c$$

$$= \frac{\frac{x}{3}}{9\sqrt{1-\left(\frac{x}{3}\right)^2}} + c = \frac{x}{27\sqrt{1-\frac{x^2}{9}}} + c \boxed{= \frac{x}{9\sqrt{9-x^2}} + c}$$

Notes:

- $3^2 - 3^2\sin^2\theta = 3^2(1-\sin^2\theta) = 3^2\cos^2\theta$ because $\sin^2\theta + \cos^2\theta = 1$.
- $(3^2\cos^2\theta)^{3/2} = (3^2)^{3/2}(\cos^2\theta)^{3/2} = 3^3\cos^3\theta$ because $(x^m)^n = x^{mn}$.
- $\frac{3}{3^3} = \frac{3^1}{3^3} = 3^{1-3} = 3^{-2} = \frac{1}{3^2} = \frac{1}{9}$ because $\frac{x^m}{x^n} = x^{m-n}$.
- $\frac{\cos\theta}{\cos^3\theta} = \frac{1}{\cos^2\theta} = \sec^2\theta$ because $\frac{x}{x^3} = \frac{1}{x^2}$ and $\frac{1}{\cos\theta} = \sec\theta$.
- $\int \sec^2\theta\, d\theta = \tan\theta$ because $\frac{d}{d\theta}\tan\theta = \sec^2\theta$ (as noted in Chapter 3).
- $\sin\theta = \frac{x}{3}$ since $x = 3\sin\theta$ and $\cos\theta = \sqrt{1-\sin^2\theta}$ since $\sin^2\theta + \cos^2\theta = 1$.
- $27\sqrt{1-\frac{x^2}{9}} = (9)(3)\sqrt{1-\frac{x^2}{9}} = 9\sqrt{9}\sqrt{1-\frac{x^2}{9}} = 9\sqrt{9\left(1-\frac{x^2}{9}\right)} = 9\sqrt{9-x^2}$.

❽ $\int_{\theta=\pi/6}^{\pi/2} \frac{\cos\theta}{\sin^3\theta}\, d\theta \quad , \quad u = \sin\theta \quad , \quad du = \cos\theta\, d\theta$

$$u_1 = u\left(\frac{\pi}{6}\right) = \sin\left(\frac{\pi}{6}\right) = \sin(30°) = \frac{1}{2} \quad , \quad u_2 = u\left(\frac{\pi}{2}\right) = \sin\left(\frac{\pi}{2}\right) = \sin(90°) = 1$$

$$\int_{\theta=\pi/6}^{\pi/2} \frac{\cos\theta}{\sin^3\theta}\, d\theta = \int_{u=1/2}^{1} \frac{du}{u^3} = \int_{u=1/2}^{1} u^{-3}\, du = \left[\frac{u^{-3+1}}{-3+1}\right]_{u=\frac{1}{2}}^{1} = \left[\frac{u^{-2}}{-2}\right]_{u=\frac{1}{2}}^{1}$$

$$= -\frac{1}{2}[u^{-2}]_{u=\frac{1}{2}}^{1} = -\frac{1}{2}\left[\frac{1}{u^2}\right]_{u=\frac{1}{2}}^{1} = -\frac{1}{2}\left[\frac{1}{1^2} - \frac{1}{(1/2)^2}\right] = -\frac{1}{2}(1-4) = \boxed{\frac{3}{2}} = \boxed{1.5}$$

Note that $\frac{\pi}{6}$rad $= \frac{\pi}{6}\frac{180°}{\pi} = 30°$, $\frac{\pi}{2}$rad $= \frac{\pi}{2}\frac{180°}{\pi} = 90°$, and $\frac{1}{(1/2)^2} = 1 \div \frac{1}{4} = 1 \times \frac{4}{1} = 4$.

Chapter 15, Part A

❶ $\int x\, e^x\, dx$

$$u = x \quad , \quad dv = e^x dx$$

$$du = \frac{du}{dx} dx = \left(\frac{d}{dx} x\right) dx = 1\, dx = dx$$

$$v = \int dv = \int e^x\, dx = e^x$$

$$\int u\, dv = uv - \int v\, du \quad \rightarrow \quad \int x\, e^x\, dx = xe^x - \int e^x\, dx$$

$$= xe^x - e^x + c = \boxed{e^x(x-1) + c}$$

❷ $\int_{x=0}^{\pi/2} x \cos x\, dx$

$$u = x \quad , \quad dv = \cos x\, dx$$

$$du = \frac{du}{dx} dx = \left(\frac{d}{dx} x\right) dx = 1\, dx = dx$$

$$v = \int dv = \int \cos x\, dx = \sin x$$

$$\int_i^f u\, dv = [uv]_i^f - \int_i^f v\, du \quad \rightarrow \quad \int_{x=0}^{\pi/2} x \cos x\, dx = [x \sin x]_{x=0}^{\pi/2} - \int_{x=0}^{\pi/2} \sin x\, dx$$

$$= \frac{\pi}{2} \sin \frac{\pi}{2} - 0 \sin 0 - [-\cos x]_{x=0}^{\frac{\pi}{2}} = \frac{\pi}{2}(1) - 0 + [\cos x]_{x=0}^{\frac{\pi}{2}} = \frac{\pi}{2} + \cos \frac{\pi}{2} - \cos 0$$

$$= \frac{\pi}{2} + 0 - 1 = \boxed{\frac{\pi}{2} - 1} \approx \boxed{0.571}$$

Notes:

- $\int \sin x\, dx = -\cos x + c$.
- $\frac{\pi}{2}$ rad $= \frac{\pi}{2} \frac{180°}{\pi} = 90°$.
- $\sin \frac{\pi}{2} = \sin 90° = 1$ and $\cos \frac{\pi}{2} = \cos 90° = 0$.
- $-[-\cos x]_{x=0}^{\frac{\pi}{2}} = [\cos x]_{x=0}^{\frac{\pi}{2}}$. Two minus signs effectively make a plus sign.
- We used a calculator to get the numerical value at the end of the solution.

Chapter 15, Part B

❸ $\int \frac{\ln x}{x^2} dx$

$$u = \ln x \quad , \quad dv = \frac{dx}{x^2}$$

$$du = \frac{du}{dx} dx = \left(\frac{d}{dx} \ln x\right) dx = \frac{1}{x} dx = \frac{dx}{x}$$

$$v = \int dv = \int \frac{dx}{x^2} = \int x^{-2}\, dx = \frac{x^{-2+1}}{-2+1} = \frac{x^{-1}}{-1} = -x^{-1} = -\frac{1}{x}$$

$$\int u\, dv = uv - \int v\, du \quad \rightarrow \quad \int \frac{\ln x}{x^2} dx = (\ln x)\left(-\frac{1}{x}\right) - \int \left(-\frac{1}{x}\right) \frac{dx}{x}$$

$$= -\frac{\ln x}{x} + \int \frac{dx}{x^2} = -\frac{\ln x}{x} + \int x^{-2}\, dx = -\frac{\ln x}{x} + \frac{x^{-2+1}}{-2+1} + c = -\frac{\ln x}{x} + \frac{x^{-1}}{-1} + c$$

$$= -\frac{\ln x}{x} - x^{-1} + c = -\frac{\ln x}{x} + -\frac{1}{x} + c = \boxed{-\frac{1}{x}(\ln x + 1) + c} \quad \left(\text{factor out } -\frac{1}{x}\right)$$

❹ $\int_{x=0}^{\pi/6} \sin x \tan x\, dx$

$$u = \tan x \quad , \quad dv = \sin x\, dx$$

$$du = \frac{du}{dx} dx = \left(\frac{d}{dx} \tan x\right) dx = \sec^2 x\ dx$$

$$v = \int dv = \int \sin x\, dx = -\cos x$$

$$\int_{x=0}^{\pi/6} \sin x \tan x\, dx = [(\tan x)(-\cos x)]_{x=0}^{\pi/6} - \int_{x=0}^{\pi/6} (-\cos x) \sec^2 x\, dx$$

$$= -[\cos x \tan x]_{x=0}^{\pi/6} + \int_{x=0}^{\pi/6} \cos x \sec^2 x\, dx = -\left[\cos x \frac{\sin x}{\cos x}\right]_{x=0}^{\pi/6} + \int_{x=0}^{\pi/6} \cos x \frac{1}{\cos^2 x} dx$$

$$= -[\sin x]_{x=0}^{\pi/6} + \int_{x=0}^{\pi/6} \frac{dx}{\cos x} = -\left(\sin\frac{\pi}{6} - \sin 0\right) + \int_{x=0}^{\pi/6} \sec x\, dx$$

$$= -\left(\frac{1}{2} - 0\right) + [\ln|\sec\theta + \tan\theta|]_{x=0}^{\frac{\pi}{6}} = -\frac{1}{2} + \ln\left|\sec\frac{\pi}{6} + \tan\frac{\pi}{6}\right| - \ln|\sec 0 + \tan 0|$$

$$= -\frac{1}{2} + \ln\left|\frac{2}{\sqrt{3}} + \frac{1}{\sqrt{3}}\right| - \ln(1+0) = -\frac{1}{2} + \ln\frac{3}{\sqrt{3}} - \ln 1 = \boxed{-\frac{1}{2} + \ln\sqrt{3}} \approx \boxed{0.0493}$$

Notes:

- $\frac{d}{dx}\tan x = \sec^2 x$.
- $\sec x = \frac{1}{\cos x}$ and $\tan x = \frac{\sin x}{\cos x}$.
- $\frac{\pi}{6}$ rad $= \frac{\pi}{6}\frac{180°}{\pi} = 30°$.
- $\sec\frac{\pi}{6} = \sec 30° = \frac{1}{\cos 30°} = \frac{1}{\sqrt{3}/2} = \frac{2}{\sqrt{3}}$ and $\tan\frac{\pi}{6} = \tan 30° = \frac{1}{\sqrt{3}}$. (Note that $\frac{2}{\sqrt{3}}$ is the same as $\frac{2\sqrt{3}}{3}$ and that $\frac{1}{\sqrt{3}} = \frac{\sqrt{3}}{3}$ if you rationalize the denominators.)
- $\ln 1 = 0$.
- We used a calculator to get the numerical value at the end of the solution.

Chapter 15, Part C

❺ $\int x^2 \cos x\, dx$

$$u = x^2 \quad , \quad dv = \cos x\, dx$$

$$du = \frac{du}{dx}dx = \left(\frac{d}{dx}x^2\right)dx = 2x\, dx$$

$$v = \int dv = \int \cos x\, dx = \sin x$$

$$\int u\, dv = uv - \int v\, du \quad \to \quad \int x^2 \cos x\, dx = x^2 \sin x - \int \sin x\,(2x\, dx)$$

$$= x^2 \sin x - 2\int x \sin x\, dx$$

Now integrate by parts a second time, with a new choice for u and dv:

$$u = x \quad , \quad dv = \sin x\, dx$$

$$du = \frac{du}{dx}dx = \left(\frac{d}{dx}x\right)dx = 1\, dx = dx$$

$$v = \int dv = \int \sin x\, dx = -\cos x$$

$$x^2 \sin x - 2\int x \sin x\, dx = x^2 \sin x - 2\left(uv - \int v\, du\right) = x^2 \sin x - 2uv + 2\int v\, du$$

$$= x^2 \sin x - 2x(-\cos x) + 2\int(-\cos x)\, dx = x^2 \sin x + 2x\cos x - 2\int \cos x\, dx$$

$$= x^2 \sin x + 2x\cos x - 2\sin x + c = \boxed{(x^2 - 2)\sin x + 2x\cos x + c}$$

❻ $\int e^x \sin x \, dx$

$$u = \sin x \quad , \quad dv = e^x dx$$

$$du = \frac{du}{dx} dx = \left(\frac{d}{dx} \sin x\right) dx = \cos x \; dx$$

$$v = \int dv = \int e^x \, dx = e^x$$

$$\int e^x \sin x \, dx = (\sin x) e^x - \int e^x \cos x \, dx = e^x \sin x - \int e^x \cos x \, dx$$

Now integrate by parts a second time, with a new choice for u and dv:

$$u = \cos x \quad , \quad dv = e^x dx$$

$$du = \frac{du}{dx} dx = \left(\frac{d}{dx} \cos x\right) dx = -\sin x \; dx$$

$$v = \int dv = \int e^x \, dx = e^x$$

$$e^x \sin x - \int e^x \cos x \, dx = e^x \sin x - \left[(\cos x) e^x - \int e^x (-\sin x) \, dx\right]$$

$$= e^x \sin x - \left[e^x \cos x + \int e^x \sin x \, dx\right] = e^x \sin x - e^x \cos x - \int e^x \sin x \, dx$$

Set the original integral equal to the last equation above:

$$\int e^x \sin x \, dx = e^x \sin x - e^x \cos x - \int e^x \sin x \, dx$$

The 'trick' is to add $\int e^x \sin x \, dx$ to both sides (and add the constant of integration):

$$2 \int e^x \sin x \, dx = e^x \sin x - e^x \cos x + c$$

Divide both sides of the equation by 2.

$$\int e^x \sin x \, dx = \frac{e^x \sin x}{2} - \frac{e^x \cos x}{2} + c = \boxed{\frac{e^x}{2} (\sin x - \cos x) + c}$$

Check your answer: Take a derivative, applying the product rule.

$$\frac{d}{dx}\left[\frac{e^x}{2} (\sin x - \cos x) + c\right] = \frac{d}{dx}(fg) + \frac{d}{dx} c = \frac{d}{dx}(fg) + 0 = g \frac{df}{dx} + f \frac{dg}{dx}$$

$$= (\sin x - \cos x) \frac{d}{dx} \frac{e^x}{2} + \frac{e^x}{2} \frac{d}{dx} (\sin x - \cos x)$$

$$= (\sin x - \cos x) \left(\frac{e^x}{2}\right) + \frac{e^x}{2} (\cos x + \sin x) = \frac{e^x}{2} (\sin x - \cos x + \cos x + \sin x)$$

$$= \frac{e^x}{2} (2 \sin x) = e^x \sin x$$

Chapter 16, Part A

❶ $$\int_{x=0}^{3}\int_{y=0}^{\sqrt{x}} xy\,dx\,dy = \int_{x=0}^{3} x\left(\int_{y=0}^{\sqrt{x}} y\,dy\right)dx = \int_{x=0}^{3} x\left[\frac{y^2}{2}\right]_{y=0}^{\sqrt{x}} dx = \frac{1}{2}\int_{x=0}^{3} x[y^2]_{y=0}^{\sqrt{x}}\,dx$$

$$= \frac{1}{2}\int_{x=0}^{3} x\left[\left(\sqrt{x}\right)^2 - 0^2\right]dx = \frac{1}{2}\int_{x=0}^{3} x(x-0)\,dx = \frac{1}{2}\int_{x=0}^{3} x^2\,dx = \frac{1}{2}\left[\frac{x^3}{3}\right]_{x=0}^{3} = \frac{1}{6}[x^3]_{x=0}^{3}$$

$$= \frac{1}{6}(3^3 - 0^3) = \frac{27}{6} = \boxed{\frac{9}{2}} = \boxed{4.5}$$

❷ $$\int_{x=0}^{y^2}\int_{y=1}^{2} \frac{y^2}{\sqrt{x}}\,dx\,dy = \int_{y=1}^{2} y^2\left(\int_{x=0}^{y^2} \frac{dx}{\sqrt{x}}\right)dy = \int_{y=1}^{2} y^2\left(\int_{x=0}^{y^2} \frac{dx}{x^{1/2}}\right)dy$$

$$= \int_{y=1}^{2} y^2\left(\int_{x=0}^{y^2} x^{-1/2}\,dx\right)dy = \int_{y=1}^{2} y^2\left[\frac{x^{-1/2+1}}{-\frac{1}{2}+1}\right]_{x=0}^{y^2} dy = \int_{y=1}^{2} y^2\left[\frac{x^{1/2}}{1/2}\right]_{x=0}^{y^2} dy$$

$$= \int_{y=1}^{2} y^2\left[2x^{1/2}\right]_{x=0}^{y^2} dy = \int_{y=1}^{2} y^2\left[2\sqrt{x}\right]_{x=0}^{y^2} dy = 2\int_{y=1}^{2} y^2\left[\sqrt{x}\right]_{x=0}^{y^2} dy$$

$$= 2\int_{y=1}^{2} y^2\left(\sqrt{y^2} - \sqrt{0}\right)dy = 2\int_{y=1}^{2} y^2 y\,dy = 2\int_{y=1}^{2} y^3\,dy = 2\left[\frac{y^4}{4}\right]_{y=1}^{2}$$

$$= \frac{2}{4}[y^4]_{y=1}^{2} = \frac{1}{2}[y^4]_{y=1}^{2} = \frac{1}{2}[2^4 - 1^4] = \frac{1}{2}(16-1) = \boxed{\frac{15}{2}} = \boxed{7.5}$$

Notes:

- $-\frac{1}{2} + 1 = -\frac{1}{2} + \frac{2}{2} = \frac{-1+2}{2} = \frac{1}{2}$ (add fractions with a common denominator).
- $\frac{1}{1/2} = 1 \div \frac{1}{2} = 1 \times \frac{2}{1} = 2$ (to divide by a fraction, multiply by its reciprocal).

Chapter 16, Part B

❸ $$\int_{x=1}^{4}\int_{y=4}^{9}\frac{dxdy}{\sqrt{xy}} = \int_{x=1}^{4}\int_{y=4}^{9}\frac{dxdy}{\sqrt{x}\sqrt{y}} = \int_{x=1}^{4}\frac{1}{\sqrt{x}}\left(\int_{y=4}^{9}\frac{dy}{\sqrt{y}}\right)dx = \int_{x=1}^{4}\frac{1}{\sqrt{x}}\left(\int_{y=4}^{9}\frac{dy}{y^{1/2}}\right)dx$$

$$= \int_{x=1}^{4}\frac{1}{\sqrt{x}}\left(\int_{y=4}^{9}y^{-1/2}\,dy\right)dx = \int_{x=1}^{4}\frac{1}{\sqrt{x}}\left[\frac{y^{-1/2+1}}{-\frac{1}{2}+1}\right]_{y=4}^{9}dx = \int_{x=1}^{4}\frac{1}{\sqrt{x}}\left[\frac{y^{1/2}}{1/2}\right]_{y=4}^{9}dx$$

$$= \int_{x=1}^{4}\frac{1}{\sqrt{x}}\left[2y^{1/2}\right]_{y=4}^{9}dx = 2\int_{x=1}^{4}\frac{1}{\sqrt{x}}\left[y^{1/2}\right]_{y=4}^{9}dx = 2\int_{x=1}^{4}\frac{1}{\sqrt{x}}\left[\sqrt{y}\right]_{y=4}^{9}dx$$

$$= 2\int_{x=1}^{4}\frac{1}{\sqrt{x}}\left(\sqrt{9}-\sqrt{4}\right)dx = 2\int_{x=1}^{4}\frac{1}{\sqrt{x}}(3-2)\,dx = 2\int_{x=1}^{4}\frac{1}{\sqrt{x}}(1)\,dx = 2\int_{x=1}^{4}\frac{dx}{\sqrt{x}}$$

$$= 2\int_{x=1}^{4}\frac{dx}{x^{1/2}} = 2\int_{x=1}^{4}x^{-1/2}\,dx = 2\left[\frac{x^{-1/2+1}}{-\frac{1}{2}+1}\right]_{x=1}^{4} = 2\left[\frac{x^{1/2}}{1/2}\right]_{x=1}^{4} = 2\left[2x^{1/2}\right]_{x=1}^{4}$$

$$= 4\left[x^{1/2}\right]_{x=1}^{4} = 4\left[\sqrt{x}\right]_{x=1}^{4} = 4\left(\sqrt{4}-\sqrt{1}\right) = 4(2-1) = 4(1) = \boxed{4}$$

❹ $$\int_{x=0}^{5}\int_{y=x}^{2x}x^2y\,dx\,dy = \int_{x=0}^{5}x^2\left(\int_{y=x}^{2x}y\,dy\right)dx = \int_{x=0}^{5}x^2\left[\frac{y^2}{2}\right]_{y=x}^{2x}dx$$

$$= \frac{1}{2}\int_{x=0}^{5}x^2\left[y^2\right]_{y=x}^{2x}dx = \frac{1}{2}\int_{x=0}^{5}x\left[(2x)^2-x^2\right]dx = \frac{1}{2}\int_{x=0}^{5}x(4x^2-x^2)\,dx$$

$$= \frac{1}{2}\int_{x=0}^{5}x^2(3x^2)\,dx = \frac{3}{2}\int_{x=0}^{5}x^4\,dx = \frac{3}{2}\left[\frac{x^5}{5}\right]_{x=0}^{5} = \frac{3}{10}\left[x^5\right]_{x=0}^{5} = \frac{3}{10}(5^5-0^5)$$

$$= \frac{3(3125)}{10} = \frac{3(625)}{2} = \boxed{\frac{1875}{2}} = \boxed{937.5}$$

Note that $(2x)^2 = 2^2x^2 = 4x^2$ because $x^m x^n = x^{m+n}$.

Chapter 16, Part C

❺ $$\int_{x=0}^{y}\int_{y=0}^{2}\int_{z=0}^{x} xy^2z^3\,dx\,dy\,dz = \int_{y=0}^{2} y^2 \int_{x=0}^{y} x\left(\int_{z=0}^{x} z^3\,dz\right)dx\,dy$$

$$= \int_{y=0}^{2} y^2 \int_{x=0}^{y} x\left[\frac{z^4}{4}\right]_{z=0}^{x} dx\,dy = \frac{1}{4}\int_{y=0}^{2} y^2 \int_{x=0}^{y} x[z^4]_{z=0}^{x}\,dx\,dy$$

$$= \frac{1}{4}\int_{y=0}^{2} y^2 \int_{x=0}^{y} x(x^4 - 0^4)\,dx\,dy = \frac{1}{4}\int_{y=0}^{2} y^2 \int_{x=0}^{y} xx^4\,dx\,dy = \frac{1}{4}\int_{y=0}^{2} y^2\left(\int_{x=0}^{y} x^5\,dx\right)dy$$

$$= \frac{1}{4}\int_{y=0}^{2} y^2\left[\frac{x^6}{6}\right]_{x=0}^{y} dy = \frac{1}{24}\int_{y=0}^{2} y^2[x^6]_{x=0}^{y}\,dy = \frac{1}{24}\int_{y=0}^{2} y^2(y^6 - 0^6)\,dy$$

$$= \frac{1}{24}\int_{y=0}^{2} y^2y^6\,dy = \frac{1}{24}\int_{y=0}^{2} y^8\,dy = \frac{1}{24}\left[\frac{y^9}{9}\right]_{y=0}^{2} = \frac{1}{216}[y^9]_{y=0}^{2} = \frac{2^9 - 0^9}{216} = \frac{512}{216}$$

$$= \boxed{\frac{64}{27}} \approx \boxed{2.370}$$

❻ $$\int_{x=0}^{\sqrt{y}}\int_{y=0}^{\sqrt{z}}\int_{z=0}^{4} x^3y\,dx\,dy\,dz = \int_{z=0}^{4}\int_{y=0}^{\sqrt{z}} y\left(\int_{x=0}^{\sqrt{y}} x^3\,dx\right)dy\,dz = \int_{z=0}^{4}\int_{y=0}^{\sqrt{z}} y\left[\frac{x^4}{4}\right]_{x=0}^{\sqrt{y}} dy\,dz$$

$$= \frac{1}{4}\int_{z=0}^{4}\int_{y=0}^{\sqrt{z}} y[x^4]_{x=0}^{\sqrt{y}}\,dy\,dz = \frac{1}{4}\int_{z=0}^{4}\int_{y=0}^{\sqrt{z}} y\left[\left(\sqrt{y}\right)^4 - 0^4\right]dy\,dz = \frac{1}{4}\int_{z=0}^{4}\int_{y=0}^{\sqrt{z}} y(y^2)\,dy\,dz$$

$$= \frac{1}{4}\int_{z=0}^{4}\left(\int_{y=0}^{\sqrt{z}} y^3\,dy\right)dz = \frac{1}{4}\int_{z=0}^{4}\left[\frac{y^4}{4}\right]_{y=0}^{\sqrt{z}} dz = \frac{1}{16}\int_{z=0}^{4}[y^4]_{y=0}^{\sqrt{z}}\,dz = \frac{1}{16}\int_{z=0}^{4}\left[\left(\sqrt{z}\right)^4 - 0^4\right]dz$$

$$= \frac{1}{16}\int_{z=0}^{4} z^2\,dz = \frac{1}{16}\left[\frac{z^3}{3}\right]_{z=0}^{4} = \frac{1}{48}[z^3]_{z=0}^{4} = \frac{1}{48}(4^3 - 0) = \frac{64}{48} = \boxed{\frac{4}{3}} \approx \boxed{1.333}$$

WAS THIS BOOK HELPFUL?

A great deal of effort and thought was put into this book, such as:

- Breaking down the solutions to help make the math easier to understand.
- Careful selection of examples and problems for their instructional value.
- Full solutions to every problem in the answer key, including helpful notes.
- Every answer was independently checked by an international math guru.
- Multiple stages of proofreading, editing, and formatting.
- Beta testers provided valuable feedback.

If you appreciate the effort that went into making this book possible, there is a simple way that you could show it:

Please take a moment to post an honest review.

For example, you can review this book at Amazon.com or Barnes & Noble's website at BN.com.

Even a short review can be helpful and will be much appreciated. If you're not sure what to write, following are a few ideas, though it's best to describe what's important to you.

- How much did you learn from reading and using this workbook?
- Were the solutions at the back of the book helpful?
- Were you able to understand the solutions?
- Was it helpful to follow the examples while solving the problems?
- Would you recommend this book to others? If so, why?

Do you believe that you found a mistake? Please email the author, Chris McMullen, at greekphysics@yahoo.com to ask about it. One of two things will happen:

- You might discover that it wasn't a mistake after all and learn why.
- You might be right, in which case the author will be grateful and future readers will benefit from the correction. Everyone is human.

ABOUT THE AUTHOR

Dr. Chris McMullen has over 20 years of experience teaching university physics in California, Oklahoma, Pennsylvania, and Louisiana. Dr. McMullen is also an author of math and science workbooks. Whether in the classroom or as a writer, Dr. McMullen loves sharing knowledge and the art of motivating and engaging students.

The author earned his Ph.D. in phenomenological high-energy physics (particle physics) from Oklahoma State University in 2002. Originally from California, Chris McMullen earned his Master's degree from California State University, Northridge, where his thesis was in the field of electron spin resonance.

As a physics teacher, Dr. McMullen observed that many students lack fluency in fundamental math skills. In an effort to help students of all ages and levels master basic math skills, he published a series of math workbooks on arithmetic, fractions, long division, algebra, trigonometry, and calculus entitled *Improve Your Math Fluency*. Dr. McMullen has also published a variety of science books, including introductions to basic astronomy and chemistry concepts in addition to physics workbooks.

Author, Chris McMullen, Ph.D.

50 CHALLENGING CALCULUS PROBLEMS

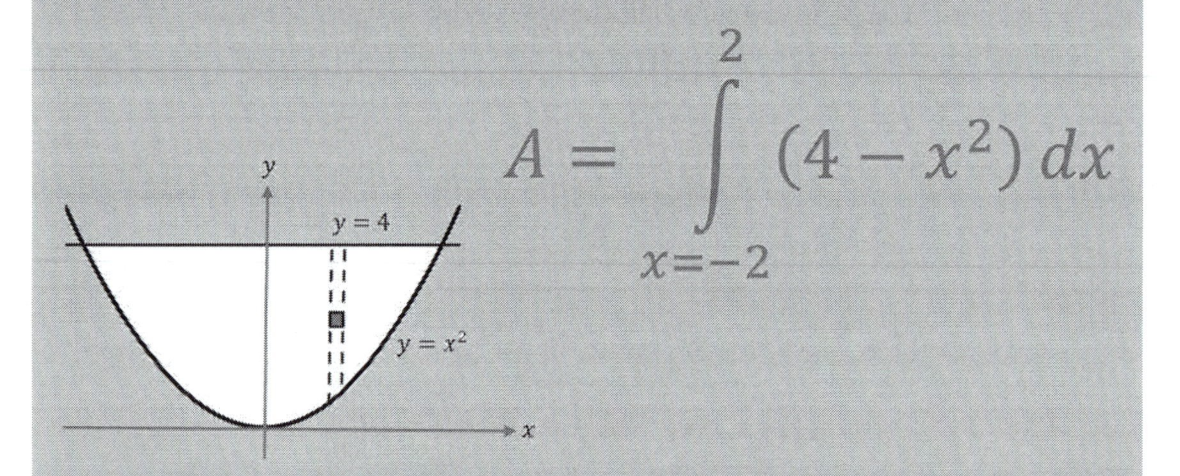

FULLY SOLVED

Chris McMullen, Ph.D.

50 CHALLENGING ALGEBRA PROBLEMS

$3x - 2y$

$9x^2 - 12xy + 4y^2$

$27x^3 - 54x^2y + 36xy^2 - 8y^3$

FULLY SOLVED

Chris McMullen, Ph.D.

Essential Calculus-based PHYSICS Study Guide Workbook

Volume 1: The Laws of Motion

$$y_{cm} = \frac{1}{m}\int y\,dm$$

$$m = \int dm = \int \sigma\,dA = \frac{\sigma \pi R^2}{2}$$

$$y_{cm} = \frac{\sigma}{m}\int_{r=0}^{R}\int_{\theta=0}^{\pi}(r\sin\theta)\,r\,dr\,d\theta$$

$$dm = \sigma dA$$

$$y_{cm} = \frac{\sigma}{m}\int_{r=0}^{R} r^2[-\cos\pi - (-\cos 0)]\,dr$$

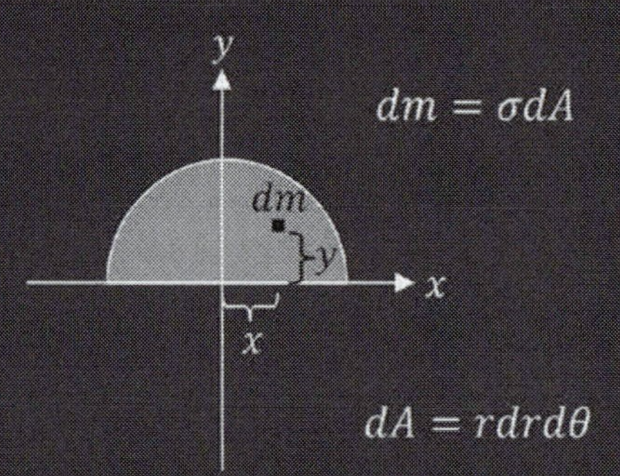

$$y_{cm} = \frac{2\sigma}{m}\int_{r=0}^{R} r^2\,dr = \frac{2\sigma R^3}{3m}$$

$$dA = r dr d\theta$$

$$y_{cm} = \frac{2\sigma R^3}{3}\frac{2}{\sigma \pi R^2} = \frac{4R}{3\pi}$$

$$y = r\sin\theta$$

Chris McMullen, Ph.D.

SCIENCE

Dr. McMullen has published a variety of **science** books, including:

- Basic astronomy concepts
- Basic chemistry concepts
- Balancing chemical reactions
- Calculus-based physics textbooks
- Calculus-based physics workbooks
- Calculus-based physics examples
- Trig-based physics workbooks
- Trig-based physics examples
- Creative physics problems

www.monkeyphysicsblog.wordpress.com

ALGEBRA

For students who need to improve their algebra skills:

- Isolating the unknown
- Quadratic equations
- Factoring
- Cross multiplying
- Systems of equations
- Straight line graphs

www.improveyourmathfluency.com

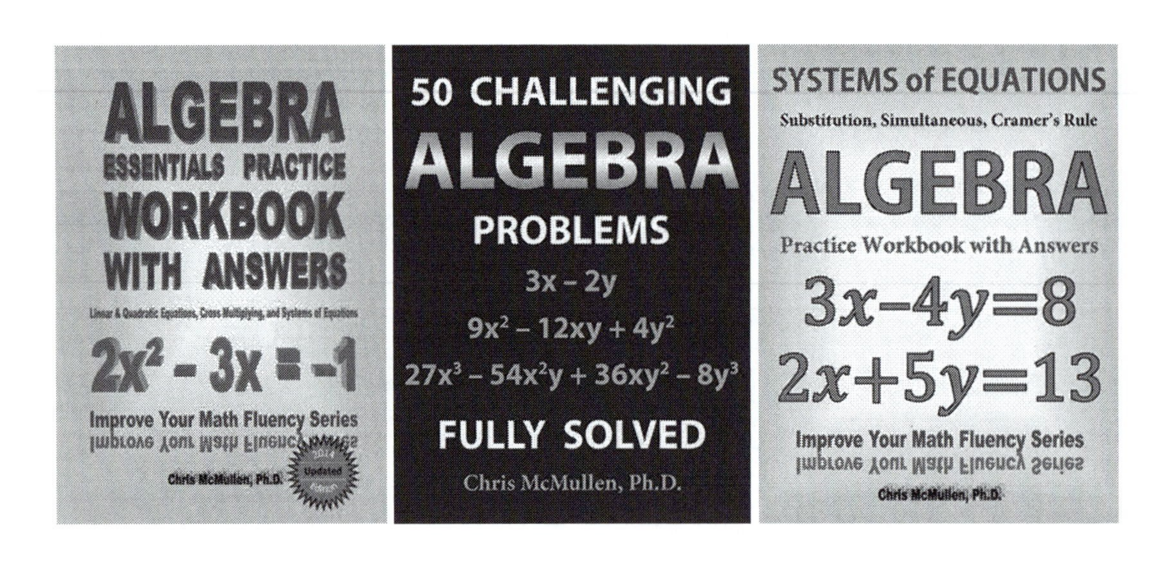